Engineering Design for Safety

Other Books of Interest

BAUMEISTER AND MARKS • *Mark's Standard Handbook for Mechanical Engineers*

BHUSHAN AND GUPTA • *Handbook of Tribology*

BRADY AND CLAUSER • *Materials Handbook*

BRALLA • *Handbook of Product Design for Manufacturing*

BRUNNER • *Handbook of Incineration Systems*

CORBITT • *Standard Handbook of Environmental Engineering*

EHRICH • *Handbook of Rotordynamics*

ELLIOT • *Standard Handbook of Powerplant Engineering*

FREEMAN • *Standard Handbook of Hazardous Waste Treatment and Disposal*

GANIC AND HICKS • *The McGraw-Hill Handbook of Essential Engineering Information and Data*

GIECK • *Engineering Formulas*

GRIMM AND ROSALER • *Handbook of HVAC Design*

HARRIS • *Handbook of Noise Control*

HARRIS AND CREDE • *Shock and Vibration Handbook*

HICKS • *Standard Handbook of Engineering Calculations*

HODSON • *Maynard's Industrial Engineering Handbook*

JONES • *Diesel Plant Operations Handbook*

JURAN AND GRYNA • *Juran's Quality Control Handbook*

KARASSIK ET AL. • *Pump Handbook*

KURTZ • *Handbook of Applied Mathematics for Engineers and Scientists*

PARMLEY • *Standard Handbook of Fastening and Joining*

ROHSENOW ET AL. • *Handbook of Heat Transfer Fundamentals*

ROHSENOW ET AL. • *Handbook of Heat Transfer Applications*

ROSALER AND RICE • *Standard Handbook of Plant Engineering*

ROTHBART • *Mechanical Design and Systems Handbook*

SHIGLEY AND MISCHKE • *Standard Handbook of Machine Design*

TOWNSEND • *Dudley's Gear Handbook*

TUMA • *Handbook of Numerical Calculations in Engineering*

TUMA • *Engineering Mathematics Handbook*

WADSWORTH • *Handbook of Statistical Methods for Engineers and Scientists*

YOUNG • *Roark's Formulas for Stress and Strain*

Engineering Design for Safety

Thomas A. Hunter, Ph.D.
Principal Consultant
Forensic Engineering Consultants, Inc.
Westpoint, Connecticut

McGraw-Hill, Inc.
New York St. Louis San Francisco Auckland Bogotá
Caracas Lisbon London Madrid Mexico Milan
Montreal New Delhi Paris San Juan São Paulo
Singapore Sydney Tokyo Toronto

Library of Congress Cataloging-in-Publication Data

Hunter, Thomas A.
Engineering design for safety / Thomas A. Hunter.
p. cm.
Includes bibliographical references and index.
ISBN 0-07-031337-7
1. Human engineering. 2. Product safety. 3. Industrial safety.
I. Title.
TA166.H86 1992
620.8′6–dc20 91-34681
CIP

1 2 3 4 5 6 7 8 9 0 DOC/DOC 9 8 7 6 5 4 3 2

ISBN 0-07-031337-7

The sponsoring editor for this book was Robert W. Hauserman, the editing supervisor was Kimberly A. Goff, and the production supervisor was Suzanne W. Babeuf. This book was set in Baskerville by McGraw-Hill's Professional Book Group composition unit.

Printed and bound by R. R. Donnelley & Sons Company.

Contents

Preface

It is my thesis that most product failures are the result of design errors. If the failure results in personal injury or property damage, then the designer is at least partially responsible for the incurred losses. The point is that designers should use the methods which are available to recognize the hazards inherent in their design, and design the hazards out of the product. It is my belief that safety should be designed into a product right from the start, rather than being tacked on after accident data have pointed out the designer's original mistakes.

Over a period of nearly ten years, I was a member of the engineering staff of a major testing and consulting organization which specialized in the technical fields. A considerable portion of our consulting work was devoted to the testing and evaluation of products that had failed in service. We were called in by the companies manufacturing these products as independent experts to determine the reason for the failure.

In many instances the failure had resulted in loss of production, imperfect or defective product, or the accidental discharge of some noxious material. While these failures did cause economic loss to the organization involved, the only thing that got hurt was the bottom line of the balance sheet.

Unfortunately, a significant number of the failures we investigated involved personal injuries and damage to property. Both people and things got hurt, usually seriously, sometimes, for the people involved in these accidents, fatally. Quite commonly the reason we were asked to provide our services was that legal actions had been initiated by the victim in an attempt to recover the losses.

The lawsuits were often brought under the law of *torts*. This means simply that someone had done something wrong and that the wrong act had caused the injury or damage to occur. On those assignments our task was not only to find out why the product failed, but also to establish who was to blame for the resulting loss. If called upon to do so, we appeared in court to testify as experts. We described the tests and examinations we carried out and gave the conclusions regarding culpability drawn from the available evidence and our reasons to substantiate those conclusions. Since some of the technical information in the case might be subject to different interpretations, the expert witness is allowed to give opinions on such matters.

Because the two sides to a lawsuit have differences of opinion which can only be resolved by a trial in court, the trial procedure is a form of debate about the evidence. The formal art of debate is called *forensics*. The debatable nature of the technical information is the reason why this branch of technology is called *forensic engineering*.

As a forensic engineer, skilled in the techniques of determining the probable cause of accidents, it has come to my attention that too many of the personal injuries and property damage losses were attributable to design errors. In too many instances, the designer was the element that caused the failure.

Most of the design failures had their basis in the lack of sufficient analysis in the initial stages of the design effort. The missing part of the analysis was in the key area of hazard recognition. Designers simply did not see that there was a hazard to life or property inherent in their approach to the problem. They did not utilize the methods that are available for conducting detailed audits of hazards, recognizing hazards that exist, identifying and classifying them, and then correcting their design mistakes by eliminating the hazard from the product right on the drawing board.

A large amount of information has been accumulated about the frequency, severity, and causes of accidents. Several volumes on this subject have been published by safety-related organizations, and additional information has been published by consumer groups and federal agencies such as the Consumer Product Safety Commission, the Department of Labor, and the Department of Defense. With this abundance of design information readily available, there is absolutely no excuse for the types of design blunders seen every day in the marketplace.

Up to now, designers have not been held personally accountable for the consequences of their errors of omission or commission. They are singularly free of professional malpractice liability. However, it is my opinion that this situation will change. My belief is that, sooner or later, legal actions will be taken against designers, just as they are now taken against other professional practitioners. And that is as it should be.

Thomas A. Hunter

Acknowledgments

The motivation to write this book came from nearly ten years of forensic engineering, involving more than 1500 projects that covered a very wide spectrum of mechanical and electrical products. Too many of these products were poorly designed, resulting in people suffering personal injuries and property suffering unnecessary damage. Technical papers on some of the worst cases have been presented at meetings of the American Society of Mechanical Engineers and reproduced in the society's monthly journal, *Mechanical Engineering*.

Robert Hauserman, a senior editor in the Professional Books Group division of McGraw-Hill, Inc., read the first of the articles and suggested that a book be prepared on the subject of designing safety into products. I am deeply grateful for his inspiration and encouragement, without them this book would never have been written. Thanks are also given to Peggy Lamb and to Kim Goff who edited the manuscript.

Much of the technical design information was obtained from United States government publications. In particular, most of the anthropometric data in the chapter on human factors were originally presented in military design handbooks, and many of the illustrations of machine-guarding methods were taken from publications of the Occupational Safety and Health Administration (OSHA). Additional human factors data were supplied by the Society of Automotive Engineers and are used with their permission.

In the chapter on "Product Liability," section 402A of the second restatement of the law of torts is quoted verbatim. This material was copyrighted in 1965 by the American Law Institute and is reprinted with their permission.

Engineering Design for Safety

1
The Concept of Safety

According to the dictionary, the word *safety* implies a condition which is free from exposure to danger, injury, or loss. When a person or thing is in a safe condition, that person or thing is out of reach of danger. For a person, being safe implies security against harm, injury, or loss to the person or property. In addition to the *condition* definition, safety is said to be a *quality* of averting or not causing injuries, danger, or loss.

Danger, on the other hand, expresses the exposure of a person or thing to harm, damage, or injury. It is a general word for liability to all kinds of losses, damages, and injuries and implies the possible existence of harm that may be encountered when someone is performing an action. As an independent word, *danger* is often used as a warning signal to alert persons to the existence of exposure to harm.

The word *safe,* by itself, implies freedom from danger. There is much debate regarding whether *safe* describes an absolute condition or a relative condition. When a product is said to be safe, is it really free from all kinds of danger under all kinds of conditions or free only from specific dangers under certain conditions? Examination of this question will quickly show that there is no such condition as being absolutely safe. There will always be a possibility for harm, even though the possibility may be minuscule and the probabilities of a given mishap may be vanishingly small. By implication, there is no such thing as an absolutely safe product. Some products may be safer than others when their respective dangers are compared. When comparable products are examined, the one which presents the least exposure to danger may be said to be *safest* even though that product, by itself, is not, and cannot be, absolutely safe. In some instances, the semantic way around this difficulty is to use the description *generally recognized as safe* (abbreviated as GRAS).

The word *hazard* is used frequently in the safety literature. From the dictionary definition, *hazard* suggests the existence of a danger which one can foresee but cannot avoid. Such terms as *hazardous waste* or *hazardous occupations* are common, implying that the dangers are known to be present, their nature is reasonably well defined, and the possible consequences of a mishap are predictable. Being unavoidable, hazards are, by this definition, something society just has to live with. Just be careful.

The dictionary definition of *hazard* is not used in what follows. In this book a hazard is considered as any sort of threat to personal safety. This applies regardless of whether the threat is foreseeable or avoidable, how well it is defined, or how well the consequences are known. In particular, *hazard* applies to threats which are not even known to exist but which turn up as a big surprise at some inopportune moment. The recognition of hazards is one of the primary themes of the design process and is discussed further later. In our view, a hazard is a hazard, regardless of its nature.

Historical Background

Folklore has it that the wheel was the first great invention; others say that the means to control fire and the ability to make and use tools were even more important. Regardless of the merits of either argument, humans have the ability to modify their environment through their own efforts. This has resulted in the design and creation of a myriad of products over the millennia. Books on the history of technology describe many of these artifacts in wondrous detail, and no repetition of this information is needed for our purposes. However, beginning with the industrial revolution and the introduction of machinery on a greatly increased scale, the presence of hazards to persons and property has increased in similar measure. The initial design efforts were focused on solving a particular problem, and the primary emphasis was on finding a workable solution to that problem. Getting the product to work at all and to perform its intended function was what mattered. Designs were worked out by trial and error, step by step, by relying on ingenuity and "animal cunning" to achieve success. Some truly remarkable machines were developed in this way, particularly in the textile industry.

Since getting the product to work was more important than other design criteria, naturally little attention was paid to the safety aspects of the design. There was little or no concern for the safety of the workers who operated the ingenious machinery. Eventually, the social costs of the number of workplace injuries became so great that legislation was introduced to regulate certain conditions of work. Compensation for injuries suffered in the workplace was assessed against the employers in the form of workers' compensation insurance. The insurance carriers, in their efforts to control the costs of this benefit program, began the development of safety improvements.

Types of Hazards

During and after the industrial revolution, the primary class of workplace hazards was in the mechanical field, with those related to the production and fabrication of hot-metal castings making a major contribution. The equipment necessary for the pouring of iron castings weighing several hundred pounds was developed about 1760, but the machinery to shape and finish these castings was already available. The steam engine, first used to remove water from underground mines, had been invented by Thomas Newcomen in the early 1700s, and the process for making wrought iron, which could be forged and machined into components for such engines, had been available for several years. Major mechanical workplace hazards have existed for almost 300 years.

Experience with such hazards provided the foundation for the recognition and classification efforts made by the early workers' compensation insurers in the latter part of the 19th century. Mechanical hazards were found to relate primarily to the injuries suffered by a machine operator coming into contact with moving parts of the machine in some unintended manner. These hazards still exist today and include ingoing nip and shear points, parts which fly out of the machine, rotating parts which have projecting contours that can catch clothing, and cutting or shearing actions involved in the machine's operation. All these hazards are related to the type of motion involved at the site of the hazard and what action that motion produces.

Only two types of mechanical motions are considered: rotating motion and transverse motion including reciprocating motion. Rotating motion involves rollers, gears, sprockets, grinding wheels, flywheels, shafting, and any other component which rotates about its own axis. The major hazard in rotating motion is the existence of an ingoing nip point between two related parts. If operators get any parts of their bodies or clothing, gloves, or tools close enough to the ingoing nip, such an object could be drawn into the nip and crushed between the mating parts.

Reciprocating and transverse motions are movements in a straight line, either continuous or in a back-and-forth manner. The hazard in this case arises when the worker or some part of his or her body is caught between the moving part and some stationary object, resulting in injury.

The actions produced by the machine in response to the applied motion can be classified into four types. These are dependent on what the machine does to the workpiece. If the workpiece is having its shape changed by removal of material, the action is *cutting, punching,* or *shearing*. If no material is to be removed, the change in the shape of the workpiece is produced by forming it under various *bending* actions.

The tools used to produce the cutting actions include saws, milling cutters, drills, and lathes. Punching is done by presses designed for that purpose, and shearing is performed by machines using knives of various configurations. In all these cases, the hazard involved is located at the precise point where the work is done on the workpiece. This place is known as the *point of operation.*

In addition to the purely mechanical hazards of the machinery itself, the act of using the machinery creates other nonmechanical injury-producing situations. A major nonmechanical hazard is noise. It has been known for many decades that prolonged exposure to high levels of noise will eventually lead to hearing impairment. In addition, a sudden and unexpected noise may startle a person and cause her or him to act in a dangerous manner. A semimechanical hazard is vibration, which can cause operator fatigue and may also lead to deterioration of the nerves and circulatory vessels in the fingers and hands.

Inherent in the use of machinery and mechanical devices is the availability of power to drive the machines. In the past, individual machines were driven by belts and pulleys which, in turn, were driven by a prime mover located in the factory power plant. Old-time machine shops were literally a forest of belts, pulleys, and shafting, all of which presented ingoing nip-point hazards where the belts ran onto the pulleys and a flying-object hazard if the belt slipped off the pulley or came apart. The development of reliable, low-cost electric motors which could be used to drive individual machines resulted in the replacement of the belt-and-pulley systems, but introduced the new hazard of exposure to electrical voltages high enough to cause death by electrocution.

Other hazards which were recognized many years ago included the danger of coming into contact with hot surfaces which were present during casting, forging, or other hot-metal processes or were generated by the operation of certain types of equipment. Any equipment which used steam for heating, cooking, or processing necessarily operated at temperatures high enough to burn humans. Scalding injuries from hot liquids are in the same general category.

With almost 200 years of accumulated experience regarding the existence and nature of hazards to persons and property, the problem of recognizing hazards in the design of new products and equipment should be well in hand and under control. Unfortunately, such is not the case, as proved by the enormous number of product liability lawsuits now in U.S. courts.

Hazard Recognition

For most people, hazard recognition begins at an early age. As soon as they are able to walk by themselves, very young children encounter the hazard of slipping and falling down. They learn to cope with this by experience and the "school of hard knocks," often assisted by the love and affection of an assisting parent or older sibling. Climbing comes next, and the exposure to more serious hazards begins immediately. Close supervision is required to prevent children's hands from grabbing the handles of kitchen utensils that stick out beyond the front of the stove and pulling scalding food down on top of their heads. In this instance, children have no means to recognize and avoid the hazard, and it is the responsibility of the supervising adult to recognize the hazard and avoid mishap by keeping the utensil handles out of the children's reach. Another potential hazard is the small freestanding stove often found in urban apartments with a pull down-door on the oven. If the oven door is left open and the child climbs up on it, the whole stove may tip forward, dropping the child to the floor and dumping whatever was cooking on the stove on top of the youngster. Again, the child cannot recognize the hazard, so someone else must take that responsibility. The point here is that often severe hazards may exist which are not recognized or recognizable by the endangered person. Obviously, someone else must be able to recognize the existence of the hazard, evaluate the seriousness of the threat, and take the necessary corrective action to avoid possible injury.

The proper person to recognize that a hazard exists, evaluate the potential threat, and work out a method for reducing the threat to an acceptable level is the person who has the overall responsibility for the design of the article being considered. The first steps in the design process involve the formulation of possible solutions to the design problem. As soon as the most promising solution has been chosen, and before any detailed design effort has been expended, research should be done of existing safety requirements or standards applicable to the product. This up-front information will serve as guidance for designers and help them avoid the inclusion of unnecessary hazards as the design progresses. It is the first step in the identification of hazards already recognized by others and known to exist. The evaluation of the threat posed by these known hazards in the particular design application is the next step for the designer. This should be followed by formulation of a suitable method for avoiding the threat.

Once the known hazards have been identified from a search of the background information, the designer should ferret out any additional hazards which are not spelled out in the literature. These are the "unknown" hazards which can show up later as a nasty surprise. If these hazards manage to get into the production process without being detected and become apparent only after a mishap occurs during product use, the resultant product liability suits may be very costly in terms of damage claims. Obviously, the most cost-effective way to avoid the "unknown" hazards is to make a systematic search for them in the early phases of the design effort.

The basic approach to finding the hidden hazards is similar to that used in establishing reliability estimates. Each component of each system in the product is examined on a "what if" basis. The question is posed, what happens if this part does not behave as intended? Clearly there may be several ways in which the behavior of the part deviates from the intended mode, and it is here that the skill of the designer comes into play. Each possible mode of failure must be established, and the consequences of that failure must be determined.

Once the consequences of failure for individual parts have been established, the process should be extended to include combinations of parts which have related functions. Some modes of failure may involve more than a single component or system so that two, three, or more elements may need to be examined in various combinations. The number of failure modes can escalate rapidly in complex systems, but the application of probability theory can eliminate those modes which are determined to be negligibly small.

The component failure mode analyses can be carried out in a reasonably straightforward manner to identify the hazards inherent in the product itself. Then comes the problem of identifying the hazards presented by the interface between the product and the user. These fall into the categories of hazards arising during expected use, those arising during misuse, and those arising from abuse.

The hazards presented during expected use can and should be identified as soon as a final design for the product has been chosen, by using the methods set forth above. The hazards presented by misuse and abuse are much more difficult to identify and evaluate since they are founded on the nearly unpredictable behavior of ordinary people.

Spectacular instances of product misuse abound, including children playing with matches, adults smoking cigarettes in bed, drivers playing "chicken" with their automobiles, and the strong man who picked up his rotary lawn mower and used it to trim the sides of a large hedge. The possibilities are limited only by the user's imagination, but some can be anticipated by the insightful designer. These are known as the *foreseeable* misuses of the product, and they and their hazards must be considered in the design process. The building and testing of models and prototypes of the product can often provide valuable insight into the existence of previously unsuspected hazards.

It is strongly recommended that designers prepare and keep a record of the intended uses and the foreseen, unintended misuses and abuses of the product which were examined during the hazard identification stages. In the event of a liability suit, this information can be used to refute any allegation which claims that the hazard which caused injury to some person or damage to some property was not properly foreseen by the designer. It may also assist in convincing a jury that the hazard was not foreseeable by a reasonably prudent and diligent designer.

Hazard Avoidance and Prevention

Once the hazards of a particular product have been recognized and identified, their characteristics can be defined. Are they pinch points, nip points, sharp edges, electrical shocks, or what? The definition stage permits a formal examination of the precise nature of the hazard and gives some insight into the level of its severity. It is at this point that the answers to the "what if" questions become important. The answers tell the designers what can be expected to happen to the user of the product under the conditions of the given failure mode. The evaluation of the severity factor is necessarily somewhat subjective, but if carried out in a conscientious manner, it does yield a ranking of the relative severity of the various recognized hazards. Those with the greatest relative severity are the most promising candidates for elimination.

Another measure of the magnitude of the threat presented by an identified hazard is the frequency with which the assumed malfunction can reasonably be expected to occur. Just as for the severity evaluation, the assignment of a frequency-of-occurrence number is also a subjective process. For products which have an established history, the frequency values can be used as a database for application to similar products with a good degree of confidence. However, for a truly new product or one which differs significantly from existing models, frequency numbers are strictly judgment calls which depend on the skill and experience of the designer.

The combination of the two hazard evaluation factors—frequency and severity—provides a good measure of the danger to persons or property imposed by the existence of the particular hazard. Obviously, once this overall index of danger has been established, the designer considers means or methods to reduce the level of danger on an item-by-item basis. Common sense dictates that the most dangerous items be dealt with first.

The systematic approach should begin with examination of the reason why the particular hazard exists. At this point, the question is, why is this hazard there at all? The answer leads to the next question: How can this hazard be eliminated completely? If there is no feasible way to eliminate the hazard completely, then the question becomes, what can be done to reduce either the expected frequency or severity factors? If either factor can be reduced, then the level of danger is reduced and the hazard is mitigated to some degree. In comparative terms, the product has been redesigned to make it "safer" than it was before. The emphasis in this whole process is on the possibilities for *prevention* of mishaps *by elimination* of the hazardous condition.

We emphasize strongly that hazard elimination is by far the most effective means to improve product safety. For this reason, designers should make hazard recognition and elimination a basic consideration in the early stages of product design. The sooner the hazard is recognized, the sooner a reduction or elimination method can be created. This will reduce the number of changes required to incorporate the improvement. Ideally, the hazard recognition and elimination process should be a fundamental part of the initial design effort.

Hazard avoidance is considered as a design approach to be used when the method of hazard elimination cannot be implemented. Certain hazards in certain products are more or less inherent, and while the hazards can be reduced, they cannot be eliminated entirely. A common example is the ordinary wooden match. The wood can be impregnated to reduce residual glow after the match has been blown out, the chemicals used in the head can be modified so that the match can be lighted only by striking it on a specially prepared surface, and instructions for use can be added to the package, but a match can still start an unwanted fire and cause damage to property and danger to life. By suitable improvements in the product, the hazard presented by matches has been reduced to a low enough level that society finds it acceptable. The benefits provided by the product are perceived to more than compensate for the costs associated with the prevailing hazard, so matches still continue to be readily available.

In the design of mechanical products, hazard avoidance methods usually involve the provision of some means of keeping the operator away from the location of the hazard so that injury cannot take place. This can be done either by putting guards around the hazard to provide a physical separation from the operator or by moving the hazard so far from the operator that the danger is avoided by the distance alone.

Hazards can also be avoided through the provision of suitable operating and maintenance instructions for the particular product. These may be incorporated directly onto the product in the form of signs, decals, or instruction plates, but are often presented in written form in a manual which accompanies the product. This method of hazard avoidance depends very heavily on the response of the user to the instructions and assumes that the user can and does read the manual before using the product. This assumption is not always met. For this reason, the instructional method of hazard avoidance is considered the least effective of the available methods.

Summary

The concept of product safety, as applied to the design of mechanical products, implies that designers of products have some measure of control over the magnitude of the threat which the products pose to people or property. The extent to which designers exert this control is a reflection of their concern for the public welfare. Another concern is the costs of litigation arising from injuries to persons or property caused by use of the product. There are established methods for the recognition and evaluation of various types of hazards. These should be applied as early as possible in the design effort. Large quantities of published materials describe ways and means of reducing the known threats. The preferred method of threat reduction is elimination of the hazard, when possible. The next best method is hazard avoidance through the use of guards or physical separation from the danger. The least desirable method is the publication of instructions or warnings.

2
Design Philosophy

To the mechanical engineer, the word *design* needs no definition or elaboration. It is the systematic process of working out the kinematics of any required motions; calculating the displacements, velocities, and accelerations involved in the motions; selecting suitable materials for the various components; proportioning them to withstand the calculated forces expected to be applied; and selecting the details of fit, form, and finish. The word *philosophy*, however, is much more abstract. It implies a way of looking at the design problem in order to establish a rational system of guidance to be used in the formulation and execution of the solution to the problem.

In this book, the system of guidance is the design review process. While clearly there may be many reasons for examination of a particular design (cost, tooling requirements, availability of materials, etc.), the emphasis here is primarily on the safety aspects. It is expected that special attention paid to the safety features of a given product may result in some conflicts with other parameters of the design. However, when the beneficial effects of the emphasis on safety are weighed against any possibly adverse effects on other factors in the decision process, careful consideration must be given to the potential costs of one special additional external factor—the cost of product liability litigation which may be incurred if the product causes injury to persons or damage to property.

Admittedly, such future costs are very difficult to evaluate since so little hard information is available about them. This is no excuse to ignore them, however, and the usual engineering approach of including a generous factor of "safety" should be applied when such costs are estimated.

There have been numerous reports in the public press of the inhibiting effect of product liability costs on innovation and the decreasing number of new-product introductions as a result of negative assessments of otherwise acceptable designs. The overall economic balance between product benefits and product liabilities is still being determined. However, the message to designers is clear. A careful and thorough analysis of the hazards in any proposed product is mandatory in the current business climate. The design review is the basic tool for carrying out the hazard analysis.

Design Reviews

Once the design problem has been formulated and it has been established just what the proposed product is intended to do, the first action the designer should do is to collect, review, and analyze all pertinent information regarding any statutory requirements, codes, standards, regulations, industry practices, or other applicable background data. In an established design department which has an ongoing effort in a specific product line, this information should be readily available and up to date.

Once the background information has been analyzed, the next step is to formulate the design constraints imposed by the known requirements. Examples include such factors as the limitation of the noise level produced by the machine when it is in operation, as specified under the federal Walsh-Healy act; the protection of ingoing nip points, as required by section 1910.212 of the Occupational Safety and Health Administration (OSHA) regulations; the provision of adequate clearances between certain electric components, as called out by the *National Electrical Code* (NEC); the inclusion of certain controls required by the *Boiler and Pressure Vessel Codes* if such items are used in the design; and the use of noninterchangeable threaded connections on the plumbing used for compressed gases, as specified by the Compressed Gas Association. There are many other examples familiar to anyone who has worked at the details of mechanical designs.

In what is euphemistically called "the good old days," the only design reviews which were performed were those carried out on the first production version of a newly designed machine. The machine would be run for a while, or until something broke. This was the "build 'em and bust 'em" era. It was often a long, tedious, and expensive process before the final version of the design was achieved. An acceptable configuration was often reached by some combination of repeated experiments, endless modifications, trial and error, animal cunning, and good luck.

The emphasis was on getting the machine to work, even though the reason why that particular trial worked might never be known. Analysis was minimal. The "practical" approach was used to get the job done. Know-how was the stock in trade of the successful designer. Ingenuity, insight, and perseverance were the traits of inventors and designers who had little technical knowledge, had almost no analytical skills and who worked with materials whose physical properties were not well defined. In spite of those limitations, enterprising designers managed to come up with a myriad of machines of astonishing complexity, which did the jobs they were intended to do. But they did not pay much attention to safety.

An outstanding example is the type casting machine. Invented by Ottmar Merganthaler in 1879 and developed over several years, it was first used in 1886 by a New York newspaper. An ingenious and amazingly complex machine, it revolutionized the printing industry by casting an entire line of type in one piece. This replaced the previous method of building up a line of type one character at a time, by hand, with the printer picking each piece of type from a wooden "case" and placing it in a holder. The pieces of type were reusable and had to be sorted and replaced in their proper places in the case after the printing was done, another time-consuming process. Merganthaler used brass matrices which were assembled into a line of type and used as a mold. A molten lead alloy was pumped into the mold, and when it solidified, the result was a cast metal slug which had the desired letters raised from its edge. One of his design problems was to get all the lines to come out the same length, i.e., to "justify" at the right end. An ingenious spacer, which incorporated a self-adjusting feature, solved that problem and made the machine commercially acceptable. He also invented a clever mechanical method for sorting and replacing the individual matrices in their cases after the slug had been cast.

There were several hazards connected with the operation of this machine, however. One was particularly dangerous. The operator sat in front of the machine keyboard, with the casting mechanism and the pot of hot metal nearby on his or her left-hand side. If the line being cast was not the proper length, the hot metal could squirt out of the casting area and strike the operator, causing severe burns.

An example of a simple device with a long and bloody accident record is the link-and-pin coupler used to hook cars together in the early days of railroading. It consisted of a forged iron link and a pair of pins. Each railroad car had a receptacle in the draw bar at each end of the car. The link fitted into the receptacle with the plane of the link in the horizontal position. A pin was then dropped through a hole in the top face of the receptacle, through the link, and through another hole in the bottom face of the receptacle. This secured the link to the end of the car in a simple and effective manner.

When another car was to be coupled to a given car, it was necessary for the worker to stand between the cars in order to guide the link into the receptacle of the added car and drop the second pin into position. The significant feature of this arrangement was that the second car was in motion while the worker was positioning the link and preparing to drop the second pin into place. If there was a sudden lurch during this operation, the worker's hand or arm could be caught and crushed between the ends of the two receptacles. The link-and-pin coupler earned the designation "the maker of one-armed men." Deservedly. It worked. It was simple. But it was so dangerous to use that better designs were developed which were operated from the side of the car. This design did not require the worker to get between the moving cars and thus eliminated the crushing hazard.

A more recent example of egregious hazards in a common product is the rotary-power lawn mower. Until about 1950 most homeowners used a reel-type mower that they pushed from behind. Although there were powered units for use on estates and golf courses where large areas of grass had to be cut, they were relatively expensive and not suitable for the small individual residential lawn. The reel mower did a satisfactory job of cutting grass, and it posed minimal hazards to the operator. However, a lot of effort was needed to propel it through long or wet grass.

It was long known that it was not necessary to shear the stalks between a fixed-bed knife and a rotating reel, as had been done for years. Grass could be cut satisfactorily if the blade was struck from the side at a speed high enough to shear off the stalk by impact forces alone. The high blade speed could be provided by an internal combustion engine running at about 3600 revolutions per minute (rpm), a commonly available engine operating point. In the resulting designs the engine was mounted on a stamped steel deck with the engine shaft in the vertical position. The cutting blade was mounted directly on the end of the shaft, turning at the 3600-rpm engine speed. The cut grass was expelled through a chute in the side of the steel deck. The mowers were highly successful in the marketplace. They worked well. They cut grass cleanly, and they required much less effort in tall grass than the reel type. They were also relatively low-cost since there were many manufacturers and competition was keen.

It was soon discovered, however, that wet grass tended to clog under the deck and plug up the entrance to the discharge chute. The obvious, and readily foreseeable, remedy for this condition was for the operator to reach into the chute and pull the obstruction out of the opening. If the operator left the engine running while clearing out the clump of wet grass, it was highly predictable that the blade would strike the operator's hand and inflict severe lacerations to the fingers. Many thousands of such injuries were reported within a few years following the introduction of rotary-powered mowers.

Another foreseeable hazard was that the rapidly rotating blade would strike a solid object that happened to be lying in the grass and expel it through the chute at high velocity. If someone happened to be standing in the vicinity of the discharge chute, the ejected object could inflict severe injury, even at considerable distances. Just as with the finger injuries, many thousands of flying-object injuries were incurred by innocent bystanders. The Consumer Product Safety Commission (CPSC) worked with the mower manufacturers to improve the rotary mower designs and reduce the severity of the hazards inherent in the original versions. The results of their efforts have been incorporated in a safety standard for lawn mowers which was sponsored by the Outdoor Power Equipment Institute and published by the American National Standards Institute (ANSI) as their standard B71.1.

The point of these examples is that reviews of the designs for these products could, should, and would have revealed the hazards connected with their normal and expected operations. A century ago the hazard associated with the link-and-pin coupler and the social costs incurred by the maimed workers were not sufficient cause to generate calls for legislative action to correct the situation. Today, by contrast, social consciousness dictates that a hazard of such severity is not permissible. Workers' compensation laws and product liability litigation have contributed to the change in social attitudes toward such remediable conditions. Design reviews are now an essential part of the process of recognizing that a hazard exists, defining the nature and severity of that hazard, and discovering ways to design the hazard out of the product before the product is created.

Design Review Process

Obviously, before a design can be reviewed, there has to be a design in some stage of formulation. As noted previously, before the first solution to the problem is conceived, the designer should have collected the applicable background information. The process of collecting these data will usually reveal the scope and nature of many hazards already known to be involved in a given product area. Unless they are novices or are breaking into an unfamiliar product line, competent designers will already be aware of most of the problem-causing features. This is known as *experience*. The design review process forces the experience out into the open where it can be recognized, evaluated, and factored into the design as it progresses. In this way, the design review process begins to consider the safety aspects even in the preliminary stages. The earlier this can be done, the easier it will be to implement any changes mandated by safety considerations.

Once the basic approach to the design problem has been worked out, the configuration of the product begins to take form on paper. The various component parts take up their positions relative to each other and the dynamics of the design fall into place. This is where the creative aspects of designers are called into play and where designers get their greatest satisfaction from solving problems. However, this is also the point at which any hazards which are present in the design should be recognized.

Hazard Recognition

From the safety viewpoint, hazard recognition in the earliest stages of any design is of paramount importance. Any threat to personal safety which can develop during the expected use of the product should be regarded as a hazard and recognized as such. Such threats can come from several sources. The following are examples of hazard sources.

Kinematic Hazards

Locations where components come together while moving, resulting in possible pinching, cutting, or crushing of any object caught between them, are classified as kinematic hazards. Sets of gears, belts running onto pulleys, matching rollers, shearing operations, and all sorts of stamping operations where the forming dies close on each other are all examples of kinematic hazards.

Energy Hazards

Components which store energy, such as springs, electric capacitors, counterbalancing weights, and compressed-gas receivers make up the class of potential energy hazards. They can cause injury if that stored energy is suddenly released in an unexpected manner. This hazard is especially important during servicing of the equipment. To eliminate it, the designer must establish a method and prepare a procedure for placing the machine in what is known as the *zero energy state* during the servicing mode. All sources of potential energy must have that energy drained out or blocked out of the product so that it cannot cause damage.

Components which have energy because of their motion present kinetic energy hazards. Such parts as flywheels, loom shuttles, fan blades, and many types of conveyors have parts which move at substantial velocities and have considerable mass. They therefore have large amounts of kinetic energy which is available to cause damage to any object which interferes with their motion.

Electrical Hazards

The principal electrical hazards to persons are the shock hazard, which may cause an undesirable involuntary motion, and the electrocution hazard, which may cause loss of consciousness or death. The major electrical hazard to property is from electrical faults, commonly called *short circuits,* which result in large releases of energy and damage to equipment. It is common for electrical faults to cause massive arcing, a cascade of sparks, and to emit flaming gobs of molten metal which can start fires in any combustible material nearby. Obviously, any person in the vicinity of a large electrical fault is in danger of being severely burned by the arc flame and the hot metal.

These hazards have been recognized ever since large-scale generation and distribution of electrical energy began a century ago. Fires caused by improperly installed electrical systems became so frequent that the fire insurance companies established Underwriters Laboratories (UL) for the purpose of developing suitable design, construction, and installation standards for electrical equipment. Thousands of UL standards have been published, and applicable ones should be obtained and used as reference material by equipment designers in the initial stages of product design.

As experience with electrical equipment accumulated, it became possible to formulate what amounted to good practices in design, installation, and operation of such devices. These efforts resulted in the preparation of electrical codes of various types by many administrative bodies. The codes have the force of law within their area of jurisdiction, and the designer must follow them when they apply.

The National Fire Protection Association (NFPA) prepared a code which became NFPA standard 70 and is known today as the *National Electrical Code* (NEC). This code has been widely adopted throughout the United States. It is revised at frequent intervals to keep its provisions abreast of the myriad of ongoing developments in electrical and electronic engineering. Thus it is important for the electrical designer to make certain that the applicable revision of the NEC is being used because so many changes are made. In addition to the code itself, there is the *National Electric Safety Code*. It is published by the American National Standards Institute (ANSI) as their standard C-2, and it should be used by the designer in conjunction with the basic code.

Because of the severity of the electrocution hazard, a great deal of attention has been paid to the details of grounding as one means of designing this hazard out of the product. A critical feature in the grounding concept is that the ground connection must always be made whenever the equipment is being used. If the ground connection is broken, or if there is an electrical fault in the ground, it is possible for the voltage to rise to a dangerous level at any location where the operator comes in contact with the equipment. The results of this condition can be fatal. To prevent it from occurring, it is now common practice to include a ground fault circuit interrupter (GFCI) in the supply line for the power. The GFCI acts to cut off the power by breaking the supply circuit, thus protecting the operator. It does this by sensing any current flow in the ground legs of the circuit. If the current exceeds the amount known to be dangerous to humans, the GFCI opens and cuts off the power. This action is similar to the operation of a conventional circuit breaker in the event of a short circuit.

Another means of reducing the probabilities of accidental contact with energized conductors is by incorporation of two levels of insulation in the design. This concept is known as *double insulation* and is widely used in electrically powered portable hand tools such as circular saws, drills, chain saws, and similar items.

An important feature in all such powered hand tools is the power cord and the plug used to connect to the power supply. If a secure ground connection is to be achieved, there must be a separate conductor in the power cord dedicated to the ground circuit alone. This is a third conductor in addition to the two used to bring in and return the power. It is intended to be electrically neutral at all times, hence requires no insulation. However, the plug on the power cord must have a third prong, and the receptacle must have a third opening to receive it. To prevent accidental misinsertions, the ground line uses circular parts which are not interconnectable with the flat blade shapes used for the energized conductors. The hazard of accidental misconnection is thereby designed out of the product.

Human Factors Hazards

In spite of the recent emphasis on automation and robotics, most industrial equipment and all consumer products are intended to be used by people. Mere mortals. They have a myriad of variations in intelligence, physical strength, computational ability, visual acuity, height, weight, education, and ethnic background. All these parameters must be considered by designers when they consider that their product must ultimately be put to use by human beings.

Fortunately, extensive reference materials are available regarding the various physical characteristics of the human body. Large numbers of people were examined to determine height, weight, length of reach, visual angles, strength in various positions, and other factors. These data have been tabulated in percentiles. A common requirement is that a given product shall be usable by anyone between the 5th and the 95th percentile for the factor being considered.

Even when all these data are used by the designer, the assumption is still made that the user of the product or the operator of the equipment will use or operate it in the manner intended by the designer. This assumption is often grossly erroneous. Human nature being what it is, the designer must also consider any other *possible* modes of use or operation and then evaluate which alternative modes are the most *probable*. Note that both probabilities and possibilities must be considered. Once the various probable alternate uses or operational modes have been determined, the hazards connected with those modes must be taken into account in the design.

The alternative uses considered by the designer usually comprise what are known as the *reasonably foreseeable* uses. The hazards can be recognized and designed out of these readily identifiable additional applications for the product. However, there may be other uses or operating modes which the designer does not anticipate, even after a diligent and conscientious search for them. These are the "sleepers" which can and do result in litigation. A key point in these cases often revolves on whether the specific use was foreseeable by a reasonably diligent designer.

A classic example of an unexpected mode of operation for a common product occurred when a husky fellow picked up his rotary-power lawn mower, turned it on its side, with the blade rotating in a vertical plane, and carried it alongside the hedge which bordered his grassy area, using the mower blade to trim the hedge. This was effective, yes, but extremely dangerous and certainly not a use or operating mode intended by the designer.

Misuse and Abuse

It is a truism that as long as people are involved, errors and mistakes will be made. When human errors do occur during use or operation of equipment, the chances for personal injury or damage to property are greatly increased. Some of these errors are accidental, and their consequences are difficult for any designer to anticipate, but others are deliberate on the part of the user. These errors are identifiable as misuse or abuse of the product.

A typical example of such product abuse is the fabrication and installation of thread adapters on containers for pressurized gases. The threads for tanks and fittings used on compressed gases are deliberately designed to be noninterchangeable. This is done to prevent the accidental misconnection and consequent hazardous mixing of certain gases. However, there are still adventurous, ignorant, or ill-advised people who will devise adapters to defeat the protective features of the compressed-gas codes. The results are often disastrous to both life and property and bear grim witness to the deliberate misuse of the product.

Product abuse, however, is usually a result of poor operating practices or a lack of maintenance. Poor operating practices are often the result of failure of the operator to read the operating instructions which usually accompany the product. The classic example of such neglect is the lawn mower which is driven by a small four-stroke-cycle gasoline engine. These engines are shipped dry, without any oil in the crankcase or fuel in the tank. The first time the mower is used, the operator is supposed to put the necessary oil in the crankcase before starting the engine. The instructions tell the operator to do this. If the user does not read or follow the instructions, but eagerly fills the tank with fuel and starts the engine, the almost immediate result is a seized piston and possibly a broken crankshaft.

Lack of maintenance is almost self-explanatory. The user of the equipment simply runs it into the ground, never makes repairs, or jury-rigs any items which fail in service. Such actions usually void any product warranty, but they do not necessarily avoid any product liability exposure in the event of a mishap.

Environmental Hazards

These hazards fall into two categories: internal and external. The internal hazards are adverse things that can happen to damage the product internally as a result of changes in the surrounding environment. External hazards are the adverse effects which the product can have to cause damage to things external to the product.

A common example of an internal environmental hazard is the effect of a drop in the temperature surrounding a water-cooled internal combustion engine. If the temperature goes below 32°F, the coolant water will freeze and the block of the engine will be damaged internally. Measures available to the designer to prevent such an occurrence include the use of so-called freeze plugs and the specification of antifreeze additives to the coolant.

It is a responsibility of the designer to determine the nature of the environment in which the product is expected to perform. In many government procurement documents, the requirements are spelled out in detail, but industrial and commercial products may not be so well defined. The designer will have to do some research to get the necessary information.

Among the factors to be considered are the extremes of temperature, the presence of any vibrations where the product is to be located, any atmospheric contaminants, the level of illumination available to the operator, ambient noise levels, and sometimes the level of electromagnetic radiation. Depending on the product, there may also be other applicable factors which have internal effects on the operation of the product.

External environmental hazards, which involve what the product does to the external world during its service life, have been receiving ever-greater attention in recent times. As examples, noise limitations have been in effect for many years to reduce the incidence of hearing loss by machine operators. Exhaust products from internal combustion engines are mandated to contain less of the nitrogen oxides and hydrocarbons. Chlorinated fluorocarbons, used extensively in refrigeration systems, are scheduled to be removed from production because of their adverse effect on the ozone layer. Polychlorinated biphenyls (PCBs), long used in electrical equipment because of their excellent dielectric properties, are now prohibited because of their severe adverse environmental effects.

Aside from chemical products, other factors which affect the environment include the thermal pollution attributed to heat exchangers, malodorous effluents from paper mills, and a wide variety of heavy-metal waste products generated during manufacturing processes. Vibrations generated by forging and stamping operations can have adverse environmental effects on the neighboring area. Electric switching devices can radiate electromagnetic disturbances which cause static in radio receivers and poor reception on TV sets. Even nearby aircraft can cause deleterious distortions of TV images by the reflections of distorted signals from their metal structural members.

The disposal of the product at the end of its useful life must also be considered within the environmental envelope. The method of disposal must be considered by the designer, and the impact of this disposal method on the environment must be assessed. As a grim example, not too many years ago designers gave no thought to how worn-out household refrigerators would be treated after they gave out. Because of the high cost of hauling away such a heavy and bulky item, many decrepit iceboxes were simply set outside. Soon it became apparent that these old units were attractive to young children. Kids would crawl into the box to hide and close the doors after them. The tragic part was that the doors latched only from the outside and could not be opened from the inside. After a number of youngsters suffocated, laws were passed requiring that the latches be removed from the doors of such refrigerators. In addition, new designs were developed that used magnetic door seals. The doors could be pushed open from inside, thus eliminating the hazard entirely.

The point is that the designer must consider the hazards posed by the proposed product throughout its entire life cycle. This must include hazards which occur during the process of making the product, the hazards which occur during the expected use of the product, the hazards which occur during foreseeable misuse and abuse of the product, hazards occurring during the servicing of the product, and the hazards connected with the disposal of the product after it has worn out.

Other Types of Reviews

While the emphasis in this book is on the use of the design review process as a means of recognizing the hazards connected with a particular product, we realize that there are other purposes for the review effort. It is common practice for there to be periodic reviews of progress, considerations given to the methods of manufacture, tooling requirements, vendor selections, make-or-buy decisions, methods of marketing the product, and input regarding competitors' efforts in the same field. Cost reduction and quality assurance requirements also get attention during design reviews. In addition, product planners will be making and revising projections of profitability as the development process continues. Perhaps the entire project may have to be terminated if some of the reviews reach negative conclusions. The weeding out of unsuitable products before large expenditures have been committed to them is one of the major benefits of the design review process.

It is important to distinguish between the design review process and the various programs and procedures for monitoring the progress of a project. Gantt charts, milestone diagrams, and critical-path methods are ways of planning a project and keeping track of its progress. They include design reviews among their component tasks but are not design reviews by themselves.

Hazard Elimination

Once the hazards in a product have been recognized through the design review process, the next step is to get rid of the hazards insofar as it is possible to do. The process of elimination involves ranking each recognized hazard in terms of its severity (the amount of property damage or personal injury which may result from a mishap) and its frequency (how often the mishap may be expected to occur). Obviously, the most severe hazards must be addressed first.

There are three generally accepted methods for removing hazards from a product. The first and most desirable method is to design the hazard out of the product. The second method is to provide guards for the hazards which cannot be designed out. The third and least desirable method is to provide warnings or instructions to alert the user or operator of the product to the fact that a hazard exists.

Because it is the most effective method, efforts to design a recognized hazard out of a product should be initiated immediately after the hazard has been rated according to its severity. This should happen as early in the design process as possible. At that time the nature of the hazard is known, and a proposed method for its elimination can be incorporated in the evolving design with minimal difficulty and least cost. Ideally, this work should be done in the preliminary design stage, when the changes can be made on paper. The next best time is when the first prototypes or mock-ups are fabricated from released drawings. The three-dimensional nature of an actual product often provides insights which were not available from the drawings, but changes are more expensive to enact at this stage.

The poorest time to make design changes for safety improvement and hazard elimination is after the product has been released to manufacturing and has been put on the market. The discovery of a design deficiency at that stage can be ruinous, particularly if the discovery is brought about by an accident which was caused by the deficiency. This situation almost inevitably results in a suit's being brought against the manufacturer by the injured party. It may also result in the recalling of the product from the marketplace. To protect themselves against massive losses, most manufacturers purchase product liability coverage, to pass the risk of loss to an insurance company. The premiums charged by insurance carriers must reflect the level of hazard they perceive to be present in the product. It follows, therefore, that effort expended to design hazards out of a product will be reflected in lower product liability insurance costs.

Testing samples of the product prior to its release to production will often reveal design deficiencies and latent hazards which were not recognized during the initial design phases. Proper and adequate testing under all the expected conditions of operation is mandatory, particularly for consumer products. One of the most common charges brought by plaintiffs' attorneys is that the product involved in the lawsuit was never adequately tested, thus permitting an unreasonably dangerous product to be sold to the public. Obviously the remedy for this is to carry out a thorough program of testing and to keep good records of the results.

Endurance testing, in which the product is operated until it fails in service, is the only way to find out if there is any additional hazard posed by the mode of failure. Although expensive and time-consuming, endurance tests are mandatory for all types of power equipment. Failures caused by fatigue of certain component parts may result in pieces being thrown out of the product in a violent manner. Serious injury to nearby persons is an obvious hazard to be recognized and designed out of the product.

Hazards attributable to excessive noise, vibration, localized high temperatures, electromagnetic radiation, or noxious emissions are most readily recognized and evaluated during a testing program. Even though the hazard may have been recognized and defined in the design review process, evaluation of the designers' attempts to eliminate or reduce the severity of the hazard can be done only by testing.

Guards and Guarding

While many recognizable hazards can be designed out of a product, not all can be successfully eliminated in that manner. A common example is the hazard presented by the closing of the dies in a stamping operation. For the machine to do its job of forming the workpiece, the two halves of the die set must come together with sufficient force to change the shape of the material in a permanent manner. If any part of the operator's body gets between the closing dies, it, too, will be permanently deformed. To prevent such an occurrence, guards must be provided which will prevent any part of the operator's body from getting between the dies during their closing motion. This guarding action can be accomplished in a variety of ways.

Guards for machinery can be classified into three types: fixed guards, interlocked guards, and adjustable guards. The adjustable guards can be further broken down into manually adjustable and self-adjusting types. All these protective features are attached to the machine itself, but can be removed when the equipment is being serviced or changed over.

An excellent description of the various types of guards, along with illustrations of different applications, is given in *Concepts and Techniques of Machine Safeguarding*. This publication was prepared by the Occupational Safety and Health Administration (OSHA) as their document 3067. It is available from the Superintendent of Documents at the Government Printing Office in Washington, D.C. Additional OSHA publications include Pamphlet 2247, *Machine Guarding*; Bulletin 2057, *Principles and Techniques of Machine Guarding*; and Pamphlet 2281, *Beware of Machine Hazards*.

Additional information is available from the National Safety Council in Chicago. Their publications include *Accident Prevention Manual for Industrial Operations*; *Guards Illustrated*; and the *Power Press Safety Manual*. The American National Standards Institute (ANSI) in New York publishes a series of safety requirements for a wide variety of machinery and equipment, including power presses, grinding wheels, lathes, power saws, press brakes, shears, automatic screw and chucking machines, die-casting machines, cold headers, and milling, drilling, and boring machines. Operator safety is the major consideration in those standards, and guarding is just one method included in their requirements.

There are several handbooks which include material on machine guarding, including *Mechanical Press Handbook* and *Press Brake and Shear Handbook,* published by Cahners Publishing in Boston, and *Standard Handbook for Mechanical Engineers,* published by McGraw-Hill in New York.

With the large amount of reference material now available on the subject of guarding, the designer of mechanical equipment usually has several design approach options for the provision of operator safety. In U.S. society today it is simply unacceptable for the designer to omit such features from the product. If an operator incurs an injury which could have been prevented by the incorporation of a suitable guard on the equipment, the machine is considered to be unreasonably dangerous. Because it is unreasonably dangerous, the machine is defective in its design, and the company which built the machine can be found liable for the injury.

Guards are not the only way to ensure the safety of machinery operators. Since the basic principle of safety is to make it impossible for any part of the operator's body to be in the danger zone in the vicinity of the recognized hazard, one obvious way to do this is to separate the operator from the hazardous zone by sufficient distance. This can be done by placing the machine controls so far away from the hazard that it is impossible for the operator to work the controls with any part of her or his body exposed to the hazard. This method of ensuring operator safety is known as *safeguarding by location*.

Another commonly used method is to provide certain devices which are auxiliary to the machinery and which may or may not be attached to it. For machines such as power presses, which have working components with a reciprocating motion, that motion can be transmitted to the worker's hands through cables which are attached to wrist cuffs. When the press makes a stroke, the movement of the ram pulls on the cables, and the cables pull the cuffs and the operator's hands out of the hazardous zone. These active restraint devices, while effective, are not popular with the press operators who feel tied to the job. Inactive restraints, which simply limit the movement of the operator's hands to prevent access to the danger zone, can be used on many types of machinery as well as on those having reciprocating parts. However, inactive restraints are sometimes resented by operators who may yield to the temptation to work without them.

One type of safeguard formerly used on presses merits special mention because it posed a hazard of its own. This was the *sweep guard.* It consisted of a stout stick which had its upper end secured to a pivot. The pivot permitted the stick to swing in a vertical arc across the operator's work area when the ram of the press came down. The idea was that the stick would knock the operator's hands and arms out of harm's way. The only trouble was that operators could be injured by the swinging stick if they did not get out of its way quickly enough. Although widely used for several years, such devices have been outlawed since more acceptable devices have become available.

Ingenious ways of controlling the power to machines have also been used to ensure operator safety. These have involved presence sensors, safety controls, and several types of interlocks. Presence sensors are devices which set up some condition sensitive to the presence of the human body. If this condition is altered, the change is detected by the sensors and used to generate a signal which prevents the machine from operating. For example, since the human body is opaque to the passage of light, a light beam and photoelectric cell can be used to scan the operator's work station. If the light beam is interrupted by any portion of the operator's body, the photocell detects this and sends a signal to the controls to stop the machine from moving.

The human body also possesses a certain amount of electrical capacitance, and this property can be used for presence detection. An electromagnetic field is established in the working area of the machine and tuned to resonate at a certain frequency. This frequency depends, among other things, on the capacitance of the area being protected. If some portion of the human body intrudes into the established field, the added capacitance of the body changes the resonant frequency of the field and detunes it. The resulting change can be detected and used to prevent the machine from operating.

Various mechanical sensors can be used such as a probe or feeler. If the operator touches the feeler, the feeler trips a small electric switch which prevents the machine from operating. Some feelers are made in the form of a pressure bar or rod. These are often placed in front of an ingoing nip-point hazard, such as exists at feed rollers. If the operator touches the bar while feeding material into the rollers, the bar senses this. A switch is actuated to shut off the machine immediately.

Machine operating controls can be designed and arranged to ensure that the operator is out of the hazardous zone. In the design of controls for power presses, e.g., it is common to have the working stroke initiated by the pressing of a pushbutton. Obviously, if only one pushbutton is provided, the operator could have one hand on the button while the other hand was in the danger zone. To prevent this, two pushbuttons are provided. Their functions are arranged in series so that both must be pressed at the same time for the machine to operate. This ensures that both of the operator's hands are out of the danger zone when the machine begins to move. When the stroke time for the machine is long enough to permit the operators to get their hands into the danger zone during the stroke, an additional refinement is incorporated. This requires that both buttons be held down continuously until the stroke is nearly completed. If either button is released during the early portions of the stroke, the motion stops immediately.

With two pushbuttons required for machine operation, some operators are tempted to tie one of the buttons down so they will have one hand free while they operate the machine with the other hand. Obviously this defeats the purpose of the two-handed safety feature. To prevent this, the controls are arranged so that both buttons must be pressed within a short time interval. If the allowable time delay is exceeded, the machine will not operate.

Interlocks are frequently used on movable guards to ensure that the guards are in place when the machine is in motion. These are usually in the form of electric switches which must be in the closed position for operating power to be supplied to the machine. If the guard is left open, the switch is also open and the machine will not operate because the power has been cut off. If the guard is opened while the machine is operating, the machine is stopped at once. While they are usually very effective, interlocks can be bypassed rather easily by simply inserting some object to keep the switch in the closed position. In addition, some machines possess considerable mechanical inertia which makes them keep on running even after the power is cut off. In those cases a powerful brake must be included on the machine to bring it to a stop quickly enough to avoid injury to the operator.

Machine Malfunctions

In the designing of guards, it is usually assumed that the machine will be operating in its normal manner. However this assumption depends upon the reliability of the machine, and it may not be justified. The clutches on heavy stamping presses occasionally fail to release at the end of the cycle, and the press will cycle again without any input from the operator. This is called a *repeat* by the machine. It is one of the most feared occurrences in any manufacturing shop. While the frequency of a repeat may be extremely low, any injury suffered by an operator who gets caught by the unexpected descent of the press ram can be extremely severe. One of the responsibilities of the machine designer is to consider the possible ways in which the machine can fail, assess the consequences of the failure, and provide a suitable means to protect the operator if such failure occurs.

Several types of failures must be considered in addition to the repeated cycle. Among the more common ones are parts which work loose in service and fall from the machine. Such falling objects can cause injury to the operator or damage to the machine if they happen to land in the working mechanism. Sometimes a malfunction in the feeding system will cause one of the workpieces to be thrown from the machine in a violent manner, or a broken drive chain may thrash about as a whip. There are other failure modes which depend on the particular machine being considered, but the underlying principle is the thorough examination of the possibilities and probabilities during the design review process. In addition, in its developmental stages the product should be tested with the specific objective of determining failure modes and consequences.

Location of Hazards

There are three common locations for hazards in machines: the point of operation, the power transmission system, and all other moving parts. Guards should be provided for all three classes of locations, and specific regulations published by OSHA relate to each of them.

The *point of operation* in a machine is defined as the place where the machine performs its work on the workpiece. For forming operations, the work involves changing the shape of the workpiece and the removal of excess material from the finished part. For material removal operations, the work may include milling, drilling, boring, turning, planing, cutting, and grinding.

Occupational safety and health standards have been established by the U.S. Department of Labor and are published in part 29 of the *Code of Federal Regulations.* Section 1910 of part 29 sets forth the OSHA regulations for general industry, and subpart O of section 1910 deals specifically with machinery and machine guarding. Subpart O includes section 1910.212. This is a general section which is very commonly cited and which gives the requirements for *all* machines. The use of the word *all* should be especially noted, since this portion is a blanket regulation which is very widely applicable. Point-of-operation guarding is treated in detail in paragraph 1910.212(a)(3)(ii). It states:

> (ii) The point of operation of machines whose operation exposes an employee to injury shall be guarded. The guarding device shall be in conformity with any appropriate standards therefor, or, in the absence of applicable specific standards, shall be so designed and constructed as to prevent the operator from having any part of his body in the danger zone during the operating cycle.

Special attention should be given to the phrases *shall be guarded* and *prevent the operator.* The use of the word *shall* makes the requirement *mandatory,* not optional. The word *prevent* requires that it be *impossible* for the operator to get any part of her or his body into the danger zone during the operating cycle. In other words, the point of operation guard *must* be designed to make it *impossible* for the operator to get into the danger zone while the machine is doing its work.

In addition to the general requirements for all machines, as set forth in paragraph 1910.212, there are OSHA requirements which apply only to specific types of machinery. These include woodworking machinery (section 1910.213), abrasive wheel machinery (section 1910.215), mills and calendars as used in the rubber and plastics industries (section 1910.216), a very detailed consideration of mechanical power presses (section 1910.217), and forging machines (section 1910.218).

The regulations for guarding of mechanical power transmission apparatus, which is the second of the three common locations for hazards, are given in section 1910.219. Much attention is paid to belts and pulleys, but chain drives, sprockets, clutches, collars, couplings, and shafting are also covered in detail. The emphasis in these sections is on the guarding of the ingoing nip point where a belt, rope, or chain runs onto a pulley or sprocket. Catch points, such as projections from collars and couplings, are required to be designed out of the power transmission components of machinery. Many OSHA requirements are based on the *Safety Code for Mechanical Power Transmission Apparatus,* which is published by ANSI as their standard B15.1. This document should be a basic reference for the designer of such equipment.

There is no special treatment by OSHA of the third location of hazards—other moving parts. Designers are expected to apply their own intellect in recognizing hazards from such sources during the performance of a hazard analysis. If such hazards cannot be designed out of the product, suitable guards must be provided if it is feasible to do so.

Portable power tools comprise an additional class of equipment which is not considered as machinery but which is very widely used in both industrial and consumer applications. There are special OSHA requirements for the guarding of such devices. They are set forth in section 1910.243, and they apply to portable power saws, belt sanders, grinders and abrasive wheel tools, explosive-actuated fastening tools, and power lawn mowers. Many of the guarding requirements are based on the information contained in standards published by ANSI for these particular tools. The standards include the *Safety Code for Woodworking Machines,* ANSI O 1.1; *Safety Code for the Use, Care, and Protection of Abrasive Wheels,* ANSI B 7.1; *Explosive Actuated Fastening Tools,* ANSI A 10.3; and *Safety Specifications for Power Lawn Mowers,* ANSI B 71. Designers must familiarize themselves with these publications as part of their background studies in the preliminary stages of a product design program.

In addition to classes and types of machinery and equipment, OSHA publishes guarding requirements which apply to entire industries. These are covered in subpart R, special industries. Just as in the case of the portable power tools, many of the requirements are derived from published ANSI standards. Section 1910.261 applies to pulp, paper, and paperboard mills. It refers to ANSI standard P 1.1, *Safety Standard for Pulp, Paper and Paperboard Mills,* and to more than 25 other such standards which are incorporated by reference. Textile machinery and equipment is regulated by section 1910.262, which draws on ANSI L 1.1, the *Textile Safety Code.* Requirements for bakery equipment are given in section 1910.263. They refer to ANSI Z 50.1, *Safety Code for Bakery Equipment.* Laundry equipment and operations are treated in section 1910.264, with references to ANSI Z 8.1, *Safety Code for Laundry Machinery and Operations.* Special requirements for sawmills and a variety of woodworking equipment are given in section 1910.265, and pulpwood logging operations are covered in section 1910.266. Both make references to several ANSI standards including O 2.1, *Safety Requirements for Sawmills,* and O 3.1, *Pulpwood Logging Safety Standards.*

Clearly, there is an abundance of information on the subject of guarding. The types of hazards which require protection by means of guards have been defined and located. The styles of guards which have been found suitable for various installations have been designed and are widely available. In addition, many varieties of protective devices have been developed which have proved highly effective. Under these circumstances, today there is no possible valid excuse for the designer of machinery to fail to provide adequate protection to machinery operators where guards are feasible.

Warnings

As noted previously, sometimes a recognized hazard cannot be designed out of the product or removed by use of a guard by any feasible means. In those situations the last, and weakest, solution to the problem is to warn the operator of the existence of the hazard. Because this condition arises so frequently, a great deal of attention has been given to the subject of warnings. Significantly, in product liability litigation, one common charge made against defendants is that they failed to provide adequate warnings to the user of a product which was claimed to have caused personal injury or property damage. Additionally, it may be claimed that the instructions for the use or operation of the product were inadequate. Designers must realize that products which lack adequate warnings or proper instructions can be legally held to be defective, even though they perform their intended function in an exemplary manner. The existence of this defect, when proved, results in liability for any damages incurred because of it.

There is a fundamental question about warnings: When are they necessary? The engineers, from their point of view, consider this as a problem of *need* rather than requirement. They usually consider the provision of a warning as an option which can be included or omitted as they see fit. If they decide that it is needed, it goes in; otherwise, not. The decision is a subjective one. The legal people, on the other hand, are concerned with the *duty* to warn, which they treat as an obligation that cannot be delegated to anyone else.

There are several criteria which designers can use for assessing the requirement for a warning. First is the listing of the intended and reasonably foreseeable uses of the product. Second is the determination of the nature of the hazards connected with each use. Third is the possibility of injury to persons or of damage to property because of an accident involving the hazard. Fourth is the probability that such an accident will take place. Fifth is the severity of the resulting injury or damage. All these factors should already have been considered in the design review process. When they are combined in one overall estimate, the result can be interpreted as a rough quantitative assessment of the overall exposure. The designer is then in a position to make the decision regarding the necessity for a warning. Again, however, in spite of what the legal profession thinks, the decision is a subjective one. At present there are no well-established go/no go limits on the value of overall exposure. The documentation prepared by the designers during the determination of the overall exposure should be preserved for the record. It can be used to defend the validity of the designer's decision in the event of subsequent legal problems.

If the designer decides to include a warning, care must be taken to ensure that the warning is adequate to achieve its intended purpose. So what is its intended purpose? First, the warning should alert the user of the product or operator of the equipment to the fact that a hazardous condition exists; second, it should explain how to avoid the danger which the hazard presents. The warning should indicate both *what* the hazard is and *how* to avoid it. The intent is to modify the behavior of the people exposed to the hazard so they will be able to avoid bodily injury or damage to property.

And how do designers ensure that the warnings are "adequate" to modify someone's behavior? Fortunately there have been many investigations seeking the answer to that question. Great relevance is placed on the human factors involved, i.e., how various people respond to certain types of warnings.

Human Factors

The results of the human factors research have shown that an effective warning has several characteristics:

1. It must attract the person's attention *immediately*.
2. It must be strong enough to be *clear and convincing*.
3. It must show how to *avoid exposure*.

To attract immediate attention, the warning must be placed where it will be easily seen in connection with the related hazard, and it must contain graphic symbols which will alert the user. Warnings found in instruction manuals are usually completely ineffective because the machine operator may never get the opportunity to read the manual. Similarly, verbal warnings given to workers by their supervisors are quickly forgotten. There is no substitute for the prominent warning located as close to the hazard as possible. It is always there, when and where it is needed.

The choice of graphic symbols to attract attention usually involves the choice of a key word or sign and of the colors in which they will be portrayed. The usual key words are *Caution, Warning,* and *Danger*. The word *Danger* is regarded as most forceful and exciting. It is used when it is 100 percent certain that exposure to the hazard will result in severe property damage or severe or fatal injury to the user. *Warning* is used when the probability of severe injury or damage is less than 100 percent, but the possibility for severe injury is still present. The word *Caution* is recommended for use in cases where the probability of injury is less than 100 percent, the possibility for severe or fatal injury is absent, but minor injury is probable.

The color used for the attention-getting words is also important. Red is regarded as the most effective, followed by orange, yellow, and black. Colors such as blue, green, and purple are not associated with hazardous conditions and should not be used on warnings. Since most warnings have a background color for the key word, the combinations of those colors must be considered. Black on white or yellow is highly visible and has excellent readability, but red on white is only fair, and red on yellow and red on green are poor choices which should not be used.

Specifications for accident prevention signs are given in the OSHA regulations in section 1910.145. By reference they incorporate two ANSI standards. Standard Z 35.1 is entitled *Specifications for Accident Prevention Signs*, and Z 53.1 is *Safety Color Code for Marking Physical Hazards*. These references should be used by designers as soon as the decision to incorporate warnings has been made.

In addition to color, the shape of a warning sign on a piece of equipment or of a warning label on a product influences its ability to attract attention. Shapes with rounded or softly curved boundaries have less attention-getting value than shapes with sharp corners. For that reason, circular, oval, or oddly curved shapes should be avoided. Rectangles are regarded as the preferred shape since they are distinctive and allow space for information to be displayed legibly. Triangles and squares are the next best choice after rectangles. Warning signs or labels with five or more sides are seldom used on industrial equipment or consumer products.

As stated previously, warnings serve two functions. The first is to alert someone to the existence of a hazard; the second is to inform the user how to avoid the danger presented by the hazard. The alerting function is taken care of by the size, shape, and colors selected for the warning. The second function requires a message. This message must be clear and convincing in order to be effective enough to meet the criterion of adequacy. Words or pictographs or a combination of the two is used to present the information. For the message to be both clear and convincing, the wording or pictographs must be chosen carefully. It is recommended that samples of several possible messages be made up and tested for their effectiveness before a final selection is made.

For a message to be clear, it must be in a language familiar to the person reading it. For articles manufactured and used entirely within one country or area which uses one common language, the language choice is obvious. But something shipped to a place where a different language is spoken will require warning messages to be translated into that language. For that reason, designers must be familiar with the linguistic characteristics of the market for the product. When the product is to be marketed in many different countries, the problem of providing warnings in several different languages can be forbidding. As more and more companies have increased the international distribution of their products, the use of pictographs, which use universal symbols, has become much more common.

If words are to be used on warnings, they should be short, and as few words as possible should be used. Long, involved sentences which use technical terms are not effective for the average person. The words must also be legible by that average person, who may have some uncorrected visual impairment. That circumstance calls for boldface type of large size and plenty of space between lines.

Pictographs have the advantage of communicating an entire idea in one symbol. They have long been used to promote highway safety and have been found to be effective even in situations where the vehicle operator has only a short time frame in which to see the sign, grasp its meaning, and take the necessary action. Pictographs are becoming more common on automotive instrument panels since the ideas they display are universally recognized. The necessity for making different panels with local languages for each country is avoided.

Pictographs for industrial equipment are less commonly used. Since they may be applied to many distinct situations involving many different types of hazards, each pictograph should be designed on an individual basis to suit the particular hazard in the particular machine. Ideally, several pictographs should be designed and tested for effectiveness before a choice is made. Only the most general guidelines are available for pictograph design.

1. Use a simple design for the symbol.
2. Express only one idea.
3. Use correct colors and shapes.
4. Place the symbol close to the words of the warning.

For consumer products, particularly powered tools and equipment, the design of suitable pictographs may be a major undertaking. The wide spectrum of intelligence, literacy, and skill levels in the consumer market necessitates rigorous testing of pictographs to ensure their effectiveness. The use of design specialists is recommended.

For all types of products, discretion should be exercised regarding the number of signs, labels, and warnings applied to a given article. Too many admonitions tend to confuse the person they are intended to help. Testing of the product, with all the warnings in place, will usually reveal the existence of any conflicts, ambiguities, or confusion. Remember that the warning must be clear and convincing in order to be effective. Testing will affirm or deny the effectiveness of a warning better than any other method.

The materials used for the warnings and how they are attached to the product are also important to the designer. Obviously a warning should not fade out or fall off before the product has reached the end of its service life. Fortunately, good decals are available which have a base of tough, wear-resistant material and a strong adhesive backing. They have been used successfully in many applications. However, in some cases a stamped or embossed plate is used. It is permanently secured to the product by rivets. Enameled panels are often used where the sign is exposed to the weather.

Instructions

While warnings are intended to alert the users of a product to the existence of a hazard and to tell them how to avoid being exposed to it, instructions are intended to inform the user of the proper way to use the product. A common example is the instruction *Shake well before using* found on many consumable items. Such instructions are worded in a positive sense, telling the user what to do, and are not necessarily related to hazards. Other instructions are provided to tell the user what *not* to do. They are usually related to some undesirable outcome if the user does the wrong thing with the product, and they may relate directly to a hazard. An example is the instruction on power lawn mowers: *Do NOT place hands or feet under deck.* Such negatively worded instructions do not let the user know the nature of the consequences of failing to follow the instruction. They have been criticized for this omission. A suggested modification of the mower instruction might be as follows: *Do NOT place hands or feet under deck. Your fingers or toes will be cut off.* While it is informative and commands the user's attention, as well as being true, it carries such strong negative connotations that mower manufacturers are understandably reluctant to use such wordings.

Detailed operating instructions for products or equipment often cannot be placed directly on the products themselves. An instruction manual or booklet is therefore usually included with the product to provide the necessary information. In addition to telling the user how to operate the equipment, the manual will often include certain warning statements in the text material. Such warnings, while an indication of the manufacturer's good intentions, are the most ineffective of all the methods available to the designer. They are not located on the product, but are filed away once they have been read, and are easily forgotten. For industrial machinery, it is a cliché that the shop supervisor has the manual in the desk drawer. The machine operator never gets to read it and may not even be aware that the manual exists.

Summary

Conducting design reviews is the recommended method of ensuring that suitable attention is given to the safety features of any product. The emphasis is on making certain that safety is designed into the product from the very beginning. This approach is in contrast to the method of designing hazards out of a product after it has been developed.

Design reviews should be initiated as soon as the product concept has been formulated. Once the function of the product has been established and the means of achieving the functional requirements have been devised, designers must begin the hazard analysis process. The hazard analysis will reveal the existence of hazards, define them in terms of their nature, assign probabilities to the frequency and severity of mishaps resulting from the hazards, and suggest ways of mitigating their consequences.

It is essential to collect and review applicable background information before any definitive design work is undertaken. Codes, standards, industry practices, and government regulations should all be evaluated to ensure that adequate attention is given to their provisions. Both internal and external types of hazards must be considered, with particular attention paid to the effects of the product on the environment. The entire life cycle of the product, from inception through disposal, should be included in the scope of the hazard analysis.

In order of preference, methods of hazard mitigation include elimination of the hazard by designing it out of the product, providing guards or devices to make it impossible for the user to get into the danger zone, attaching clear and convincing warnings to the hazardous locations, and providing written instructions for the safe use of the product.

All such designs, guards, warnings, and instructions should be tested under conditions replicating the expected use of the product as well as its foreseeable misuses and abuses.

3
Human Factors in Design

During the process of designing a product, the designer will necessarily consider many factors which influence the concept, details, or choice of materials for the product. Examples are cost factors, stress concentration factors, power factors, and factors of safety. All these factors are simply important elements to be considered during the analysis of the design problem and the synthesis of its solution. They establish the envelope of physical or economic parameters within which the product must be designer if it is to perform its intended function.

Quite commonly, however, designers, in their zeal to solve the often formidable physical and economic problems, forget that their wonderful invention is going to be used or operated by a mere mortal. The result is that some automobile seat belts chafe the driver's neck, some video cassette recorders defy their owners' efforts to program them, and some engine-driven power equipment makes so much noise that the operator suffers a hearing loss. Many other examples could be cited, but in each case the designer failed to pay enough attention to the limitations of the ultimate factor at the end of the line—the human factor.

Mention was made of human factors in Chapter 2. It was pointed out that many people may introduce additional hazards into a product beyond those recognized and considered by the designer. This could be done by misuse, abuse, or use of the product in an unintended manner. Another mention of human factors was made in regard to the design of suitable warnings and the writing of instructions. The emphasis in those instances was on the visual and syntactic requirements for the warnings and instructions to be effective. In this chapter, the ergonomic and other human factors aspects of design are considered.

So what is ergonomics? The first three letters of the word, erg, provide the clue. The word *erg* is derived from the Greek word *ergon*, which means work. In physics an erg is defined as the amount of work done by a force of one dyne moving through a distance of one centimeter in the direction of the force. Ergonomics, therefore, considers the ability of the user to perform physical work. It focuses on the size of the force required to perform a particular movement and the distance through which that movement is carried out. It includes the effects of fatigue on the worker when repetitive operations are performed over a certain span of time.

Some dictionaries consider ergonomics to be synonymous with human factors. However, as just mentioned, the ability to perform physical work is just one of the human parameters which the designer must take into account. There are many other factors, such as reaction times, visual perception limitations, and ability to withstand heat or cold. A broader definition of human factors includes all the limitations imposed by the body and mind of the user or operator of the product.

Sources of Human Factors Information

Fortunately, a great deal of research has been done to determine and define various human dimensions and performance limitations. The data collected from these research efforts have been compiled into tables, charts, and standards which have been published by various organizations. Among them is the Society of Automotive Engineers (SAE), which has prepared what the society calls a Recommended Practice with the title *Human Physical Dimensions*, SAE J833. This document defines the sizes of people from all over the world. The dimensions given in SAE J833 are closely correlated to the International Standards Organization (ISO) standard 3411, *Human Physical Dimensions of Operators*. It is intended to be used by designers of construction equipment, general-purpose industrial machinery, agricultural tractors, forestry machinery, and special mining equipment.

For statistical reasons, the sample of people used for the database was chosen to represent the entire spectrum of human variability, but the smallest and largest 5 percent of the sample were excluded from the recommended practice. The objective was to provide physical dimensions which would cover everyone from the 5th to the 95th percentile for both females and males. Dimensions are given for small, medium, and large humans, with small being defined as the 5th percentile female, medium as the 50th percentile of the whole sample, and large as the 95th percentile male.

As shown in Fig. 3.1, data are given for 28 different dimensions for a seated person; 42 items are shown in Fig. 3.2 for someone in a standing position. All dimensions are given in millimeters. The approximate

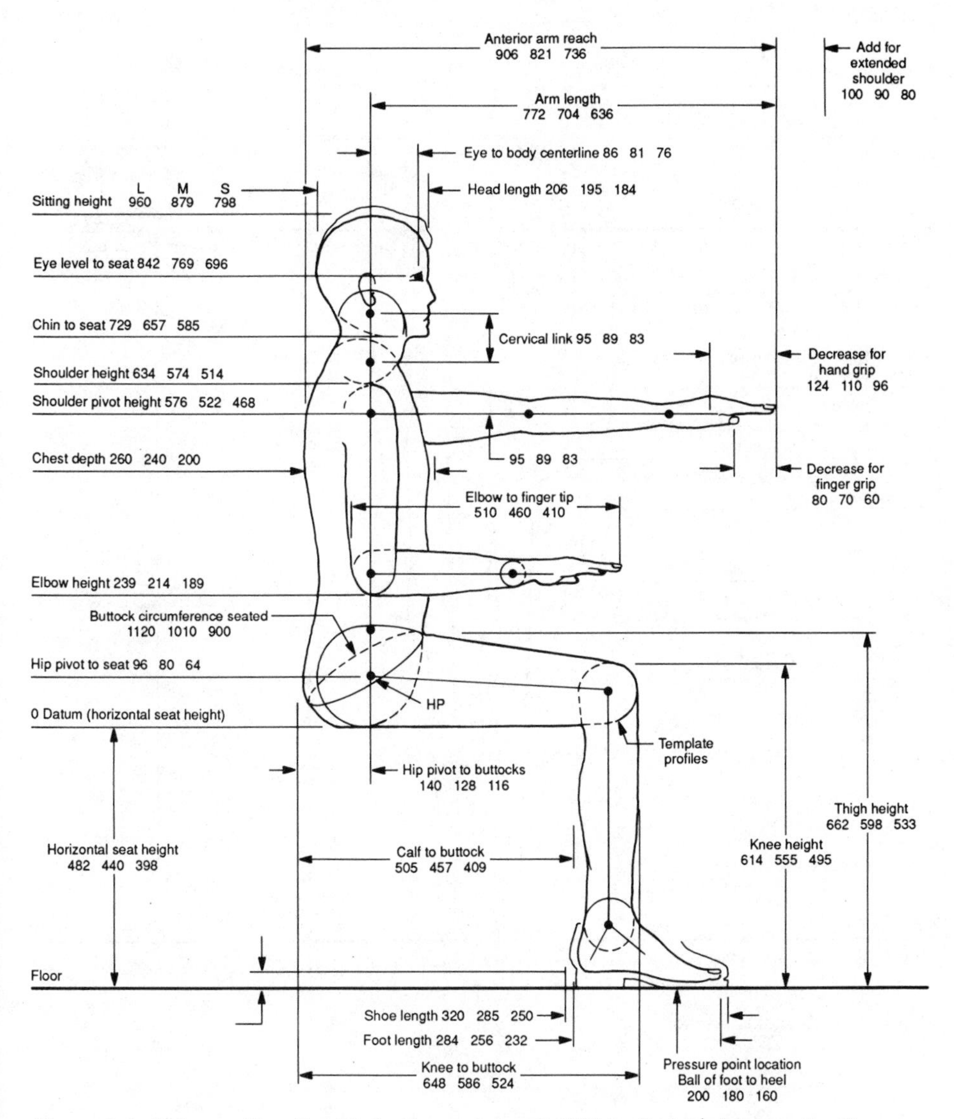

Figure 3.1. Sitting position. Reprinted with permission © 1989 by Society of Automotive Engineers, Inc.

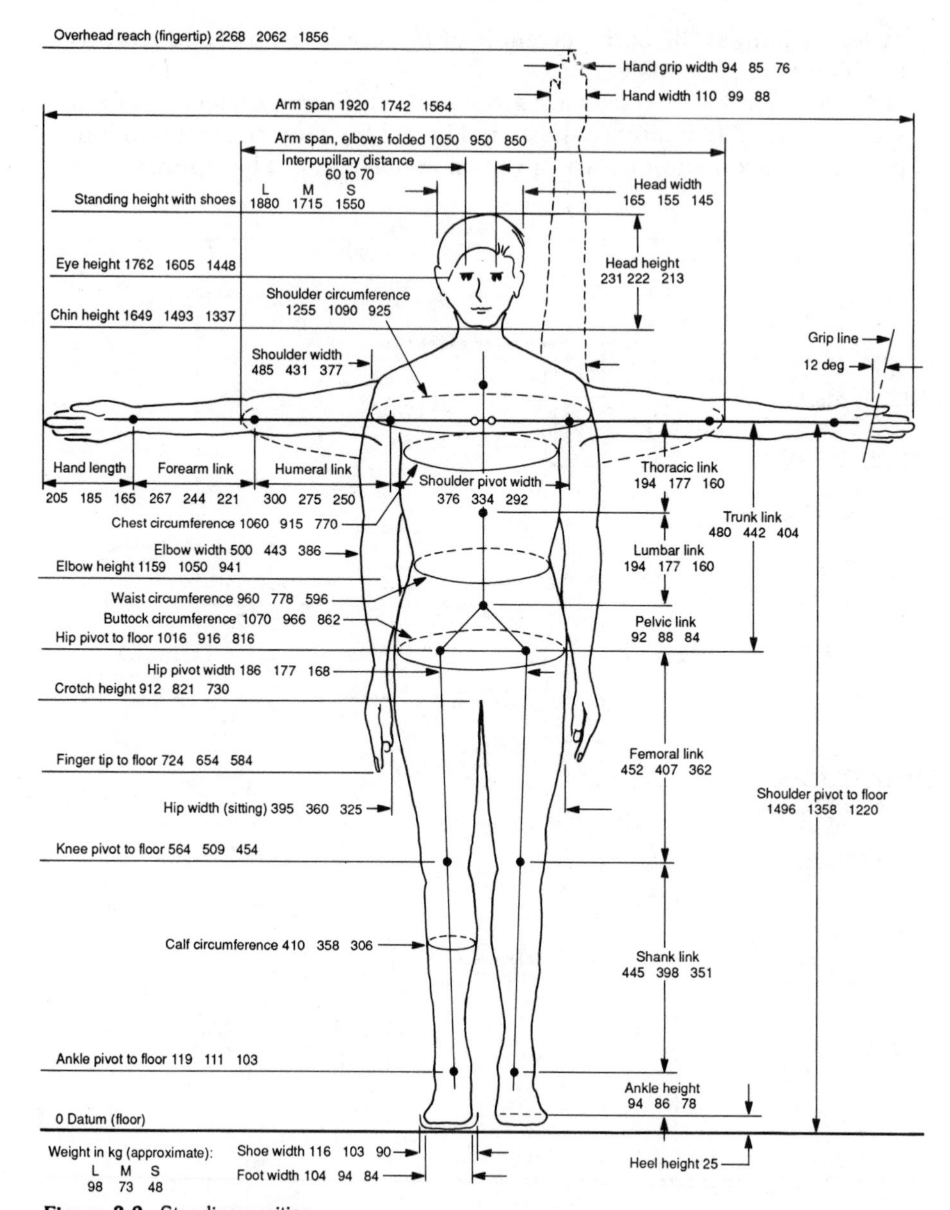

Figure 3.2. Standing position.

body weights are given in kilograms. With more than 70 specific characteristics defined in a statistically reliable manner, designers can be sure that the human body is extremely well described. The SAE data are typical of what information is available. Other readily accessible sources are also in print.

In addition to J833, SAE has published other standards relating to the human body. These include J925, which gives the minimum dimensions for service access for off-road machinery, and J1460, which is a report containing information on the dynamic response of the upper portions of the body to accelerations and inertia forces. Under the subject heading of human engineering, SAE lists 15 documents relating to design standards. These include the design criteria for the location of hand controls used by the drivers of passenger vehicles and light trucks, J1138 and J1139; shin-knee positions and stomach positions for truck drivers, J1521 and J1522; design considerations for operator enclosures on construction and industrial equipment, J154a; and the vision factors to be considered in the design of rear-view mirrors, J985. All these design-oriented documents, plus hundreds of others, are contained in the SAE handbook. This standard reference consists of a set of four volumes and is revised yearly. Copies of the handbook or individual documents can be obtained from the Society of Automotive Engineers, 400 Commonwealth Avenue, Warrendale, PA 15096.

As might be expected, the military and space agencies of the U.S. government have been active in the investigation of human factors. NASA has been particularly diligent in this field because of the exceedingly complex systems involved in space exploration. Integrating a human being into the sophisticated electronic systems was a major challenge and was researched in depth. The results of these efforts have been published as *Man-System Integration Standards* (MSIS), which is designated NASA STD 3000. It is available from the MSIS Custodian at NASA-Johnson Space Center, Houston, TX 77058. In addition to considering the physical dimensions, which are part of the SAE documents, NASA stresses the relationship of the human performance capabilities to the requirements of the system. Elaborate systems analysis techniques are included in its approach to the design problem.

The scientific term for the measurement of humans is *anthropometrics*. NASA has conducted extensive anthropometric investigations in addition to those already carried out by SAE. These data have been published as a three-volume *Anthropometrics Source Book*. Volume I is devoted to data for designers; volume II is a handbook of anthropometric data; and volume III is an annotated bibliography on the subject of anthropometrics. These data are available from the NASA Scientific and Technical Information Office, Yellow Springs, OH 45387.

Defense Department Sources

The Department of Defense (DOD) has also been deeply involved in human factors applications for many years. Because of the tremendous number of different items purchased by the military, considerable attention has been paid to standardization in order to reduce the number of variations in the system. The guidelines for designers of military materiel are published in the form of handbooks and specifications which reflect the emphasis on standardization.

The basic DOD document is *Human Engineering Procedures Guide,* the handbook designated DOD-HDBK-763. The Army, as a military department, has published *Human Factors Engineering Design for Army Materiel,* MIL-HDBK-759A, and the Air Force Systems Command (AFSC) has issued a design handbook *Human Factors Engineering,* AFSC DH 1-3. In addition there are two basic military specifications for military systems, equipment, and facilities. These are *Human Engineering Design Criteria,* MIL-H-1472D, and *Human Engineering Requirements,* MIL-H-46855B. All the DOD and MIL documents can be obtained from the Standardization Documents Order Desk, 700 Robbins Avenue, Philadelphia, PA 19111. The Air Force handbook comes from Wright-Patterson Air Force Base, Dayton, OH 45433.

The Human Engineering Requirements, as set forth in MIL-H-46855, call for the use of MIL-STD-1472 for the design criteria to be used to satisfy the requirements. STD-1472 is called for in the "Applicable Documents" section of the handbook. This is typical of the military method of referring to other background and reference materials. Although only this one specific document is listed as applicable in this instance, it is common for military specifications to list many other standards, handbooks, specifications, regulations, and publications from other military, federal, and nongovernment agencies and organizations. In this way a sizable "specification tree" is built up.

The designer should be aware that the applicable documents that are called for may, in themselves, refer to still other documents. As an example, MIL-STD-1472 refers to 25 MIL standards, 14 MIL specifications, 5 ANSI standards, 4 federal publications, 2 MIL handbooks, 2 ASTM standards, and 1 each of International Standards Organization (ISO) and SAE standards. Furthermore, the listed applicable documents also have their own list of still more documents. The tree can grow to unmanageable proportions very quickly. To avoid a document disaster, designers on military or government projects should establish the limit of the number of steps in the tree at the earliest possible opportunity. Two steps are recommended. From experience, three steps should be the absolute maximum.

The requirements section of MIL-H-46855 is quite specific in its language. Its basic purpose is to "achieve the effective integration of personnel into the design of the system." That really defines what human factors is all about, although the military uses the term *human engineering*.

The general requirements of the specification are broken into three basic sections. The first is analysis. The analysis effort begins with a definition of what the system or product is intended to accomplish. The military calls this the *mission* of the system. It answers the question, What is this thing supposed to do? Once that has been established, the mission is broken down into functions, and the functions are further divided into tasks. Each task is examined to determine what human performance parameters are applicable, what equipment will be required, and what environmental conditions will prevail while the task is being carried out. Any high-risk areas in the tasks are identified at this early stage so that proper attention can be given to them during subsequent stages of the system design. This approach is simply an example of sound human factors engineering. It can be applied to consumer, industrial, and commercial products almost without alteration.

The second of the general requirements is the design and development phase. This effort transfers the results from the analysis phase to the detail designs "to create a personnel-system interface that will operate within human performance capabilities, meet system functional requirements, and accomplish mission objectives." In nonmilitary terms, this means that the product has to be usable by the customer, work well enough to be acceptable in the marketplace, satisfy customer needs, and make a profit. Note, again, that human factors engineering is introduced in the beginning of the design and development effort. Note, too, that design reviews are an inherent part of the detail design and development effort.

The third portion of the general requirements of MIL-H-46855 is the test and evaluation program. This is intended to verify that the product, as designed and developed, meets the human factors engineering criteria established in the analysis phase and is compatible with the overall system requirements. There is certainly nothing new in that. Any commercial product should be subject to the same type of testing and for the same reasons.

As mentioned previously, one function of the analysis effort is to identify any tasks which pose high risks. These risks may be in the nature of danger to the operator, possible unsafe practices, or the requirement of critical performance on the part of the operator. All these factors should be considered by designers of commercial products. Specifically, the specification calls for the identification of the following items for each critical task to be performed by either the equipment operator or the persons who maintain the equipment:

1. Information needed to initiate the task, including necessary cues
2. Information available to perform the task
3. How the person evaluates the information
4. The decision reached after the information has been assessed
5. Action taken to accomplish the task
6. Body movement required by action
7. Workspace envelope required for action
8. Workspace available for action
9. Work environment
10. How fast and how often the action is done
11. How accurate the action must be
12. Information feedback to let the operator know how adequate his or her actions are
13. Tools and equipment required for the task
14. Number of people required for the task and their qualifications
15. Reference materials, manuals, guides, and instructions required
16. Type and quality of communications needed
17. Nature of any special hazards involved
18. Operational performance limits of personnel and equipment
19. Nature and limitations of operator interactions when more than one person is required to perform the designated task

Obviously, not all these items apply to every task, but they serve as an excellent checklist which can be applied to the design of a wide spectrum of commercial products.

Item 9, the work environment, is given special attention in MIL-H-46855. The design of the work environment, operator stations, and facilities which affect human performance capabilities must take into account many applicable factors. Among them are the following:

1. Ambient atmosphere, including odors, contaminants, temperature, humidity, air movement, and pressure in enclosed areas
2. Climate and weather in unenclosed areas
3. Acoustic noise and vibrations, including both steady and impulsive types
4. Effects of physical or emotional fatigue
5. Effects of clothing and personal protective equipment
6. Illumination

All these factors should be considered for normal operating conditions and for foreseeable abnormal and emergency situations.

Military Human Factors Design Requirements

As mentioned previously, the human engineering criteria to be used in the design of military items are given in MIL-STD-1472. This is the document called out in MIL-H-46855 as the controlling design specification. The objectives of the standard are simple, straightforward common sense which can be applied directly to the design of civilian and commercial products with equal validity. They include the following:

1. The product shall provide a suitable work environment.
2. The work environment shall promote good work patterns, safety, and health for personnel and shall foster effective operating procedures.
3. Factors which degrade human performance or increase error rates shall be minimized.
4. Operator workload, accuracy, time constraints, thought processes, and communication requirements shall not exceed operator capabilities.
5. Design shall minimize training requirements for operator and maintenance personnel.

Note that all these wonderful features are to be *designed into* the product from the beginning of the design effort, *not* tacked on at the end after deficiencies have shown up in the tests and evaluations program. This same design philosophy is entirely applicable to nonmilitary products. It should be incorporated into industrial design procedures as a matter of routine, sound engineering, and good business.

The standard gives more than 250 pages of detailed design requirements covering visual displays, audio displays, controls, labeling, the workspace, ambient environment, and maintenance. Special attention is devoted to the design of small articles and portable equipment which is intended to be carried from place to place. Vehicles intended for use on shipboard or on the ground are especially scrutinized with regard to their design features because they are so common in the military services.

Hazards and safety are treated in a separate section of MIL-STD-1472. However, most of the stated design requirements are of a general nature and follow other published safety practices. Examples are the guards required to protect personnel from the hazards of moving machinery or power transmission equipment. Guarding is generally in conformity with the OSHA regulations and the recommendations of the National Safety Council. Electrical hazards are mentioned only briefly, and designers are advised to refer to the *National Electrical Code* for detailed information on this subject. As discussed later in this chapter, MIL-HDBK-759A gives much more detailed attention to the human factors aspects of safety.

Thermal contact hazards, mentioned only occasionally in the usual literature, must be prevented by guards for both high- and low-temperature exposures. The type of exposed surface is separated into metal, glass, and wood or plastic because of the widely different thermal conductivities of those materials. Obviously more heat will be transferred to a person's skin from a highly conducting metal surface than from a poor conductor such as wood. Therefore the allowable temperature limits for a given exposure time will be higher for poor conductors than for good ones. The specified temperature limits beyond which guards are required are as follows:

Exposure	Metal	Glass	Wood or plastic
Momentary	32 to 140°F	32 to 154°F	32 to 185°F
Prolonged	32 to 120°F	32 to 138°F	32 to 156°F

Burn criteria for human skin are shown in Fig. 3.3.

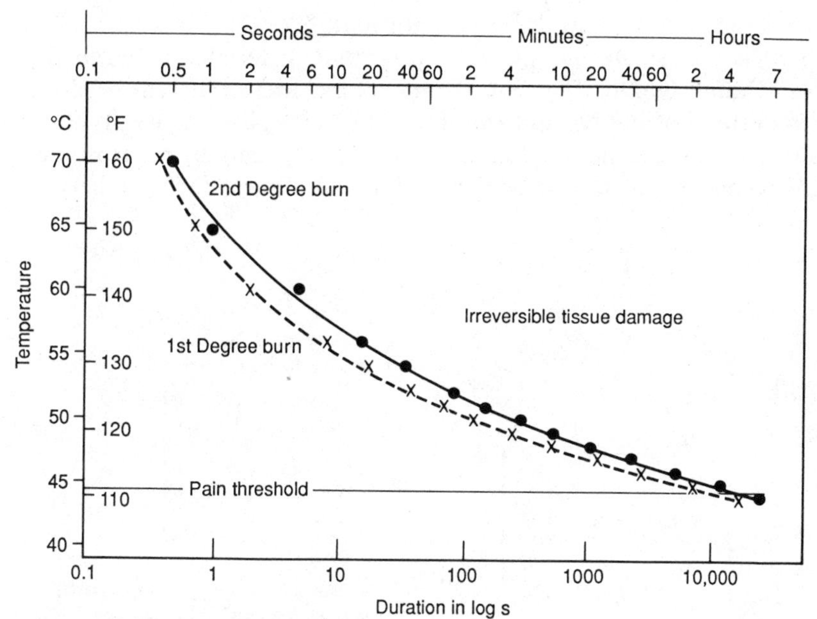

Figure 3.3. Burn criteria for human skin.

Anthropometric Data

With such large numbers of people to draw from, the military has had a wonderful opportunity to gather statistically valid data on the physical dimensions of human adults. The type of information is very similar to that presented in SAE J833, *Human Physical Dimensions,* but is much more detailed. A total of 91 different dimensions are tabulated, with the data given in both metric and English units.

The sample of persons used for the data included only those who could pass the military physical examinations. Hence the data are drawn from a somewhat narrower spectrum of the general population than the SAE data. However, the differences caused by this selection process are considered insignificant for the designers' purposes. Both datasets span the population from the 5th to 95th percentiles, thereby including 90 percent of the total group. Instead of classifying people as small, medium, or large, as in the SAE data, the military breaks the classes into ones convenient for their use: ground troops, aviators, and women.

The sample groups used were quite large. The ground troop sample included nearly 9000 men from both the Army and the Marine Corps; the aviators were chosen from nearly 5500 flying personnel of the Army, Navy, and Air Force; the data for women were taken from 1300 Army and 1905 Air Force personnel, including nurses in both instances. For information and reference, the data are shown in

Figs. 3.4 through 3.9 and Tables 3.1 through 3.7.

There is one small, but possibly important difference between the SAE and military data. The SAE numbers include the height of shoes and the effects of lightweight clothing. They also give a table of allowance to be made for heavy clothing. The military data, by contrast, are measurements of unclothed bodies, with no allowances for shoe height.

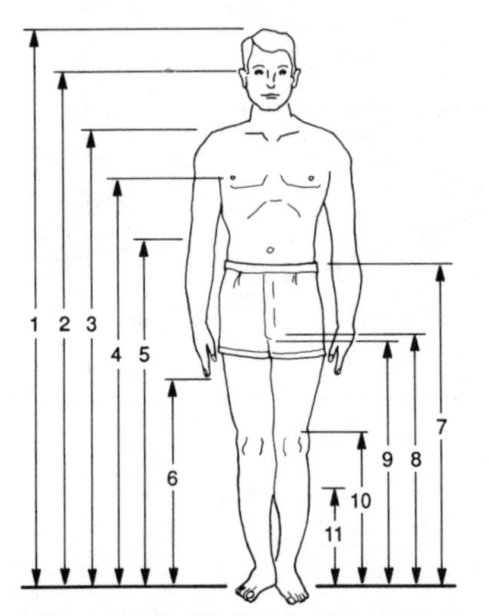

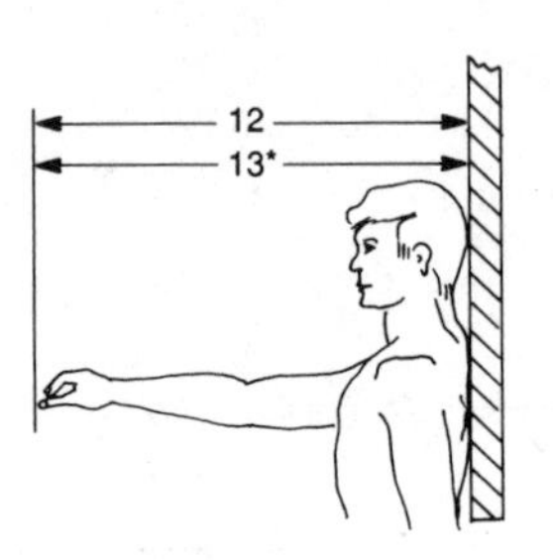

*Same as 12; however, right shoulder is extended as far forward as possible while keeping the back of the left shoulder firmly against the back wall.

Figure 3.4. Standing body dimensions.

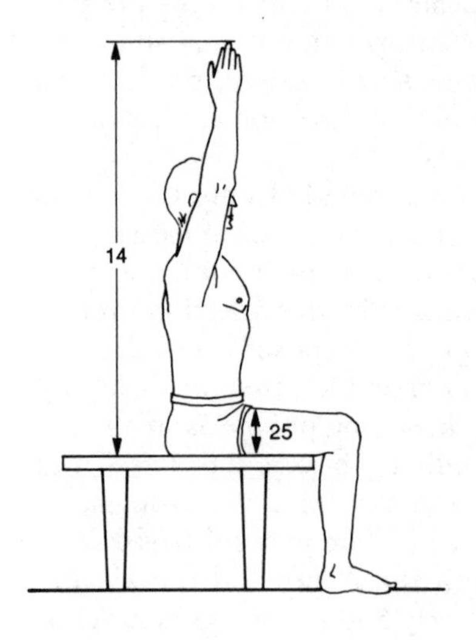

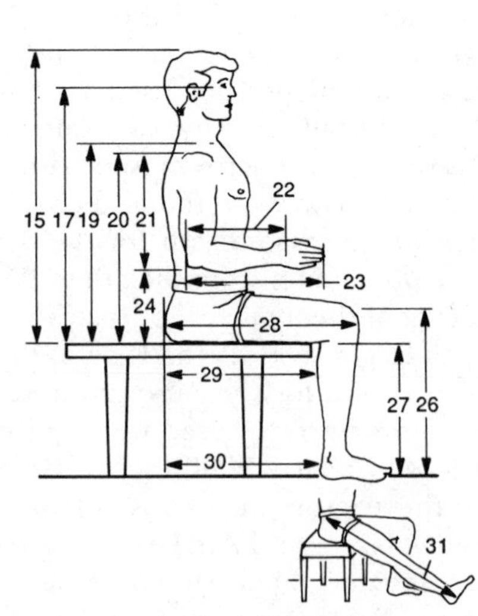

Figure 3.5. Seated body dimensions.

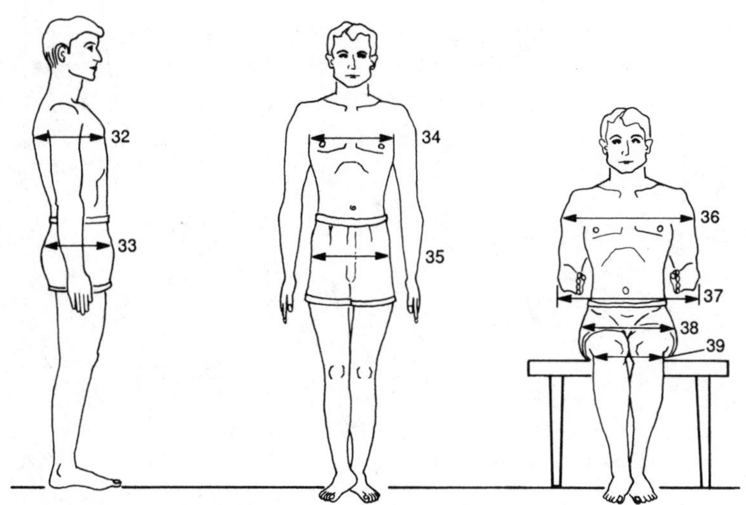

Figure 3.6. Depth and breadth dimensions.

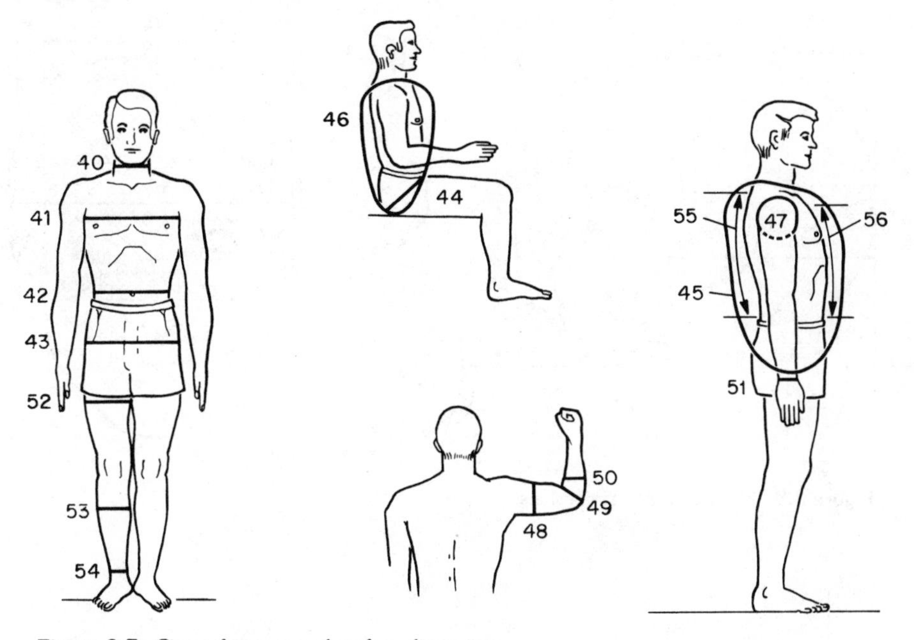

Figure 3.7. Circumferences and surface dimensions.

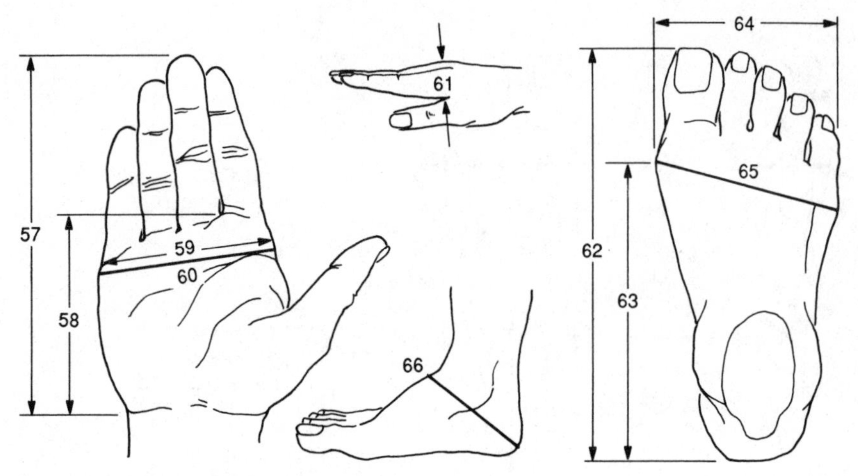

Figure 3.8. Hand and foot dimensions.

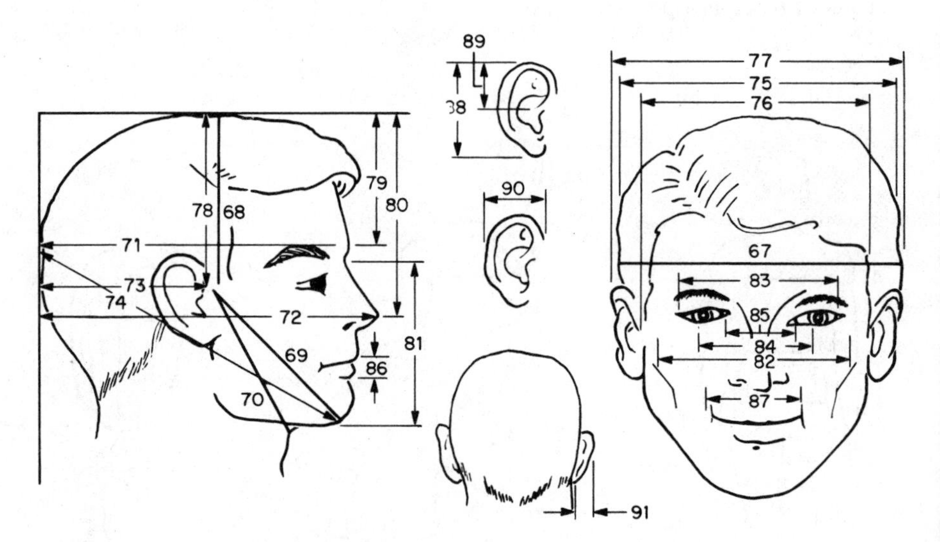

Figure 3.9. Head and face dimensions.

Table 3.1. Standing Body Dimensions

Standing Body Dimensions	5th Percentile Ground troops	5th Percentile Aviators	5th Percentile Women	95th Percentile Ground troops	95th Percentile Aviators	95th Percentile Women
	Percentile values in centimeters					
Weight (kg)	55.5	60.4	46.4	91.6	96.0	74.5
1. Stature	162.8	164.2	152.4	185.6	187.7	174.1
2. Eye height (standing)	151.1	152.1	140.9	173.3	175.2	162.2
3. Shoulder (acromiale) height	133.6	133.3	123.0	154.2	154.8	143.7
4. Chest (nipple) height*	117.9	120.8	109.3	136.5	138.5	127.8
5. Elbow (radiale) height)	101.0	104.8	94.9	117.8	120.0	110.7
6. Fingertip (dactylion) height		61.5			73.2	
7. Waist height	96.6	97.6	93.1	115.2	115.1	110.3
8. Crotch height	76.3	74.7	68.1	91.8	92.0	83.9
9. Gluteal furrow height	73.3	74.6	66.4	87.7	88.1	81.0
10. Kneecap height	47.5	46.8	43.8	58.6	57.8	52.5
11. Calf height	31.1	30.9	29.0	40.6	39.3	36.6
12. Functional reach	72.6	73.1	64.0	90.9	87.0	80.4
13. Functional reach, extended	84.2	82.3	73.5	101.2	97.3	92.7
	Percentile values in inches					
Weight (lb)	122.4	133.1	102.3	201.9	211.6	164.3
1. Stature	64.1	64.6	60.0	73.1	73.9	68.5
2. Eye height (standing	59.5	59.9	55.5	68.2	69.0	63.9
3. Shoulder (acromiale) height	52.6	52.5	48.4	60.7	60.9	56.6
4. Chest (nipple) height*	46.4	47.5	43.0	53.7	54.5	50.3
5. Elbow (radiale) height	39.8	41.3	37.4	46.4	47.2	43.6
6. Fingertip (dactylion) height		24.2			28.8	
7. Waist height	38.0	38.4	36.6	45.3	45.3	43.4
8. Crotch height	30.0	29.4	26.8	36.1	36.2	33.0
9. Gluteal furrow height	28.8	29.4	26.2	34.5	34.7	31.9
10. Kneecap height	18.7	18.4	17.2	23.1	22.8	20.7
11. Calf height	12.2	12.2	11.4	16.0	15.5	14.4
12. Functional reach	28.6	28.8	25.2	35.8	34.3	31.7
13. Functional reach, extended	33.2	32.4	28.9	39.8	38.3	36.5

*Bust height for women

SOURCE: MIL-STD-1472D

Table 3.2. Seated Body Dimensions

Seated body dimensions	5th Percentile Ground troops	5th Percentile Aviators	5th Percentile Women	95th Percentile Ground troops	95th Percentile Aviators	95th Percentile Women
	Percentile values in centimeters					
14. Vertical arm reach, sitting	128.6	134.0	117.4	147.8	153.2	139.4
15. Sitting height, erect	83.5	85.7	79.0	96.9	98.6	90.9
16. Sitting height, relaxed	81.5	83.6	77.5	94.8	96.5	89.7
17. Eye height, sitting erect	72.0	73.6	67.7	84.6	86.1	79.1
18. Eye height, sitting relaxed	70.0	71.6	66.2	82.5	84.0	77.9
19. Midshoulder height	56.6	58.3	53.7	67.7	69.2	62.5
20. Shoulder height, sitting	54.2	54.6	49.9	65.4	85.9	60.3
21. Shoulder–elbow length	33.3	33.2	30.8	40.2	39.7	36.6
22. Elbow–grip length	31.7	32.6	29.6	38.3	37.9	35.4
23. Elbow–fingertip length	43.8	44.7	40.0	52.0	51.7	47.5
24. Elbow rest height	17.5	18.7	16.1	28.0	29.5	26.9
25. Thigh clearance height		12.4	10.4		18.8	17.5
26. Knee height, sitting	49.7	48.9	46.9	60.2	59.9	56.5
27. Popliteal height	39.7	38.4	38.0	50.0	47.7	45.7
28. Buttock–knee length	54.9	55.9	53.1	65.8	65.5	63.2
29. Buttock–popliteal length	45.8	44.9	43.4	54.5	54.6	52.6
30. Buttock–heel length		46.7			56.4	
31. Functional leg length	110.6	103.9	99.6	127.7	120.4	118.6
	Percentile values in inches					
14. Vertical arm reach, sitting	50.6	52.8	46.2	58.2	60.3	54.9
15. Sitting height, erect	32.9	33.7	31.1	38.2	38.8	36.8
16. Sitting height, relaxed	32.1	32.9	30.5	37.3	38.0	35.3
17. Eye height, sitting erect	28.3	30.0	26.6	33.3	33.9	31.2
18. Eye height, sitting relaxed	27.6	28.2	26.1	32.5	33.1	30.7
19. Midshoulder height	22.3	23.0	21.2	26.7	27.3	24.6
20. Shoulder height, sitting	21.3	21.5	19.6	25.7	25.9	23.7
21. Shoulder–elbow length	13.1	13.1	12.1	15.8	15.6	14.4
22. Elbow–grip length	12.5	12.8	11.6	15.1	14.9	14.0
23. Elbow–fingertip length	17.3	17.6	15.7	20.5	20.4	18.7
24. Elbow rest height	6.9	7.4	6.4	11.0	11.6	10.6
25. Thigh clearance height		4.9	4.1		7.4	6.9
26. Knee height, sitting	19.6	19.3	18.5	23.7	23.6	21.8
27. Popliteal height	15.6	15.1	15.0	19.7	18.8	18.0
28. Buttock–knee length	21.6	22.0	20.9	25.9	25.8	24.9
29. Buttock–popliteal length	17.9	17.7	17.1	21.5	21.5	20.7
30. Buttock–heel length		18.4			22.2	
31. Functional leg length	43.5	40.9	39.2	50.3	47.4	46.7

SOURCE: MIL-STD-1472D

Table 3.3. Depth and Breadth Dimensions

Depth and breadth dimensions	5th Percentile Ground troops	5th Percentile Aviators	5th Percentile Women	95th Percentile Ground troops	95th Percentile Aviators	95th Percentile Women
	Percentile values in centimeters					
32. Chest depth*	18.9	20.4	19.6	26.7	27.8	27.2
33. Buttock depth		20.7	18.4		27.4	24.3
34. Chest breadth	27.3	29.5	25.1	34.4	38.5	31.4
35. Hip breadth, standing	30.2	31.7	31.5	36.7	38.8	39.5
36. Shoulder (bideltoid) breadth	41.5	43.2	38.2	49.8	52.6	45.8
37. Forearm–forearm breadth	39.8	43.2	33.0	53.6	60.7	44.9
38. Hip breadth, sitting	30.7	33.3	33.0	38.4	42.4	43.9
39. Knee-to-knee breadth		19.1			25.5	
	Percentile values in inches					
32. Chest depth*	7.5	8.0	7.7	10.5	11.0	10.7
33. Buttock depth		8.2	7.2		10.8	9.6
34. Chest breadth	10.8	11.6	9.9	13.5	15.1	12.4
35. Hip breadth, standing	11.9	12.5	12.4	14.5	15.3	15.6
36. Shoulder (bideltoid) breadth	16.3	17.0	15.0	19.6	20.7	18.0
37. Forearm–forearm breadth	15.7	17.0	13.0	21.1	23.9	17.7
38. Hip breadth, sitting	12.1	13.1	13.0	15.1	16.7	17.3
39. Knee-to-knee breadth		7.5			10.0	

*Bust depth for women.

SOURCE: MIL-STD-1472D

Illumination

In the work environment factors listed in MIL-H-46855, illumination is one item over which the product designer has almost complete control. The effects of clothing, operator fatigue, and ambient atmosphere on operator performance are usually things the designer has to live with. Illumination, however, is a well-documented factor with requirements which can readily be designed into a product at the outset.

Military standard illumination requirements are given for many types of specific tasks in Table 3.8, and recommendations for display lighting are given in Table 3.9. In many situations, however, the amount of illumination required is controlled by other standards. In such instances, designers must refer to the controlling documents. These may be as diverse as local building code requirements, OSHA regulations, or various ANSI standards.

Table 3.4. Circumferences and Surface Dimensions

Circumferences	5th Percentile Ground troops	Aviators	Women	95th Percentile Ground troops	Aviators	Women
	Percentile values in centimeters					
40. Neck circumference	34.2	34.6	29.9	41.0	41.6	36.7
41. Chest circumference*	83.8	87.5	78.4	105.9	109.9	100.2
42. Waist circumference	68.4	73.5	59.5	95.9	101.7	83.5
43. Hip circumference	85.1	87.1	85.5	106.9	108.4	106.1
44. Hip circumference, sitting		97.0	87.7		119.3	110.8
45. Vertical trunk circumference; standing	150.6	156.3	142.2	178.6	181.9	166.3
46. Vertical trunk circumference, sitting		150.4	134.8		175.0	161.0
47. Arm scye circumference	39.6	39.9	33.6	50.3	53.0	41.7
48. Biceps circumference, flexed	27.0	27.8	23.2	37.0	36.9	30.8
49. Elbow circumference, flexed		28.5	23.5		34.2	30.0
50. Forearm circumference, flexed	26.1	26.3	22.2	33.1	33.1	27.5
51. Wrist circumference	15.7	15.3	13.6	18.6	19.2	16.2
52. Upper thigh circumference	48.1	49.6	48.7	63.9	66.9	64.5
53. Calf circumference	31.6	33.3	30.6	41.2	41.3	39.2
54. Ankle circumference	19.3	20.0	18.7	25.2	24.8	23.3
55. Waist back length	39.2	42.4	36.7	50.8	50.9	45.4
56. Waist front length	36.1	35.7	30.5	46.2	44.2	41.4
	Percentile values in inches					
40. Neck circumference	13.5	13.6	11.8	16.1	16.4	14.4
41. Chest circumference*	33.0	34.4	30.8	41.7	43.3	39.5
42. Waist circumference	26.9	28.9	23.4	37.8	40.0	32.9
43. Hip circumference	33.5	34.3	33.7	42.1	42.7	41.8
44. Hip circumference, sitting		38.2	34.5		47.0	43.6
45. Vertical trunk circumference, standing	59.3	61.6	56.0	70.3	71.6	65.5
46. Vertical trunk circumference, sitting		59.2	53.1		68.9	63.4
47. Arm scye circumference	15.6	15.7	13.2	19.8	20.9	16.4
48. Biceps circumference, flexed	10.6	11.0	9.1	14.6	14.5	12.1
49. Elbow circumference, flexed		11.2	9.2		13.5	11.8
50. Forearm circumference, flexed	10.3	10.4	8.7	13.0	13.0	10.8
51. Wrist circumference	6.2	6.0	5.4	7.3	7.6	6.4
52. Upper thigh circumference	18.9	19.5	19.2	25.1	26.3	25.4
53. Calf circumference	12.4	13.1	12.0	16.2	16.3	15.4
54. Ankle circumference	7.6	7.9	7.4	9.9	9.7	9.2
55. Waist back length	15.4	16.7	14.4	20.0	20.0	17.9
56. Waist front length	14.2	14.1	12.0	18.2	17.4	16.3

*Bust circumference for women.

SOURCE: MIL-STD-1472D

Table 3.5. Hand and Foot Dimensions

Hand and foot dimensions	5th Percentile Ground troops	5th Percentile Aviators	5th Percentile Women	95th Percentile Ground troops	95th Percentile Aviators	95th Percentile Women
	Percentile values in centimeters					
57. Hand length	17.4	17.7	16.1	20.7	20.7	20.0
58. Palm length	9.6	10.0	9.0	11.7	11.9	10.8
59. Hand breadth	8.1	8.2	6.9	9.7	9.7	8.5
60. Hand circumference	19.5	19.6	16.8	23.6	23.1	19.9
61. Hand thickness		2.4			3.5	
62. Foot length	24.5	24.4	22.2	29.0	29.0	26.5
63. Instep length	17.7	17.5	16.3	21.7	21.4	19.6
64. Foot breadth	9.0	9.0	8.0	10.9	11.6	9.8
65. Foot circumference	22.5	22.6	20.8	27.4	27.0	24.5
66. Heel–ankle circumference	31.3	30.7	28.5	37.0	36.3	33.3
	Percentile values in inches					
57. Hand length	6.85	6.98	6.32	8.13	8.14	7.89
58. Palm length	3.77	3.92	3.56	4.61	4.69	4.24
59. Hand breadth	3.20	3.22	2.72	3.83	3.80	3.33
60. Hand circumference	7.68	7.71	6.62	9.28	9.11	7.82
61. Hand thickness		0.95			1.37	
62. Foot length	9.65	9.62	8.74	11.41	11.42	10.42
63. Instep length	6.97	6.88	6.41	8.54	8.42	7.70
64. Foot breadth	3.53	3.54	3.16	4.29	4.58	3.84
65. Foot circumference	8.86	8.91	8.17	10.79	10.62	9.65
66. Heel–ankle circumference	12.32	12.08	11.21	14.57	14.30	13.11

SOURCE: MIL-STD-1472D

Design Requirements or Maintainability

Although the equipment operator or the user of a product will necessarily be given priority in human factors design efforts, sometime during its service life that equipment or product is probably going to need some maintenance or repair work. In many instances the type of work to be done and how often it is to be performed are spelled out in recommended maintenance procedures that accompany the product at the time of delivery. As an example, every purchaser of a new automobile receives a schedule of recommended maintenance in the owner's manual.

While one of the worthwhile objectives of product designers is to re-

Table 3.6. Head and Face Dimensions, cm

Head and face dimensions	5th Percentile			95th Percentile		
	Ground troops	Aviators	Women	Ground troops	Aviators	Women
	Percentile Values					
67. Head circumference	53.2	53.8	52.2	58.8	59.9	57.7
68. Bitragion-coronal curvature	31.9	33.4	31.3	36.1	37.8	36.3
69. Bitragion-menton curvature	29.0	30.1	27.3	33.1	34.7	31.6
70. Bitragion-submandibular curvature	26.7	28.4	24.5	30.7	33.6	28.9
71. Head length	18.2	18.6	17.3	20.7	21.0	19.8
72. Pronasale to wall	20.8	21.4	19.7	23.5	24.1	23.2
73. Tragion to wall	8.5	9.2	8.8	12.6	12.1	11.8
74. Head diagonal (menton–occiput)		24.4			26.9	
75. Head breadth	14.2	14.4	13.5	16.3	16.5	15.6
76. Bitragion breadth	12.5	13.1	12.1	14.5	15.2	13.8
77. Biauricular breadth	16.5	17.5	14.2	19.4	20.2	17.4
78. Head height (trag.-top of head)	11.9	12.0	11.6	14.5	14.4	14.3
79. Glabella to top of head	6.5	7.2	7.1	9.4	10.9	9.9
80. Pronasale to top of head	11.6	13.0	11.9	15.1	16.6	16.8
81. Face length (menton–sellion)	10.6	10.2	9.6	13.1	13.0	11.8
82. Face (bizygomatic breadth)	12.8	12.4	11.9	14.9	15.1	14.0
83. Biocular breadth	9.3	8.4	8.8	10.9	10.1	10.5
84. Interpupillary breadth	5.1	5.3	5.1	6.8	7.0	6.5
85. Interocular breadth		2.7	2.7		3.8	3.7
86. Lip-to-lip length		1.1			2.3	
87. Lip length (mouth breadth)		4.5	3.7		5.9	5.1
88. Ear length	5.5	5.9	4.5	6.9	7.3	6.0
89. Ear length above tragion		2.5			3.4	
90. Ear breadth	3.8	3.0	2.4	5.0	4.3	3.5
91. Ear protrusion		1.6			2.8	

SOURCE: MIL-STD-1472D

duce the amount of maintenance required, another obvious objective should be to make the performance of the known necessary tasks as simple and convenient as possible for the person who eventually has to do the job. MIL-STD-1472 gives detailed guidelines for human factors in equipment maintenance. Among them are several which can be applied to nonmilitary products:

1. Avoid the use of special tools.
2. Make wear-out items replaceable as modules.

Table 3.7. Head and Face Dimensions, in.

Head and face dimensions	5th Percentile			95th Percentile		
	Ground troops	Aviators	Women	Ground troops	Aviators	Women
	Percentile values					
67. Head circumference	20.94	21.18	20.57	23.16	23.59	22.73
68. Bitragion-coronal curvature	12.56	13.14	12.31	14.21	14.90	14.29
69. Bitragion-menton curvature	11.42	11.86	10.74	13.03	13.66	12.45
70. Bitragion-submandibular curvature	10.51	11.18	9.63	12.09	13.23	11.37
71. Head length	7.19	7.32	6.80	8.14	8.27	7.80
72. Pronasale to wall	8.18	8.42	7.88	9.27	9.50	9.15
73. Tragion to wall	3.33	3.62	3.47	4.95	4.77	4.64
74. Head diagonal (menton–occiput)		9.60			10.59	
75. Head breadth	5.59	5.67	5.33	6.40	6.50	6.12
76. Bitragion breadth	4.92	5.17	4.76	5.71	5.98	5.45
77. Biauricular breadth	6.50	6.89	5.61	7.64	7.95	6.84
78. Head height (trag.-top of head)	4.69	4.74	4.55	5.72	5.69	5.62
79. Glabella to top of head	2.56	2.81	2.79	3.70	4.30	3.88
80. Pronasale to top of head	4.57	5.12	4.70	5.94	6.54	6.61
81. Face length (menton–sellion)	4.17	4.04	3.79	5.17	5.13	4.63
82. Face (bizygomatic breadth)	5.04	4.87	4.69	5.88	5.94	5.53
83. Biocular breadth	3.66	3.31	3.47	4.29	3.99	4.14
84. Interpupillary breadth	2.01	2.10	2.00	2.67	2.75	2.57
85. Interocular breadth		1.08	1.05		1.50	1.45
86. Lip-to-lip length		0.41			0.92	
87. Lip length (mouth breadth)		1.76	1.46		2.30	2.01
88. Ear length	2.17	2.31	1.77	2.72	2.88	2.34
89. Ear length above tragion		0.97			1.36	
90. Ear breadth	1.50	1.19	0.95	1.97	1.70	1.38
91. Ear protrusion		0.65			1.09	

SOURCE: MIL-STD-1472D

3. Design adjustments to be mutually independent.
4. Provide failure indications to make troubleshooting fast, simple, and easy.
5. Check for proper accessibility to parts.
6. Design for goofproof reassembly (use noninterchangeable fittings, distinctive pin alignment patterns, color codes, etc.).
7. Use knobs for frequently made adjustments.

Table 3.8. Specific Task Illumination Requirements

Work area or type of task	Illumination levels, lux* (fc)	
	Recommended	Minimum
Assembly, missile component	1075 (100)	540 (50)
Assembly, general		
Coarse	540 (50)	325 (30)
Medium	810 (75)	540 (50)
Fine	1075 (100)	810 (75)
Precise	3230 (300)	2155 (200)
Bench work		
Rough	540 (50)	325 (30)
Medium	810 (75)	540 (50)
Fine	1615 (150)	1075 (100)
Extra Fine	3230 (300)	2155 (200)
Bomb shelters and mobile shelters, when used for rest and relief	20 (2)	10 (1)
Business machine operation (calculator, digital, input, etc.)	1075 (100)	540 (50)
Console surface	540 (50)	325 (30)
Corridors	215 (20)	110 (10)
Circuit diagram	1075 (100)	540 (50)
Dials	540 (50)	325 (30)
Electrical equipment testing	540 (50)	325 (30)
Emergency lighting	NA	30 (3)
Gages	540 (50)	325 (30)
Hallways	215 (20)	110 (10)
Inspection tasks, general		
Rough	540 (50)	325 (30)
Medium	1075 (100)	540 (50)
Fine	2155 (200)	1075 (100)
Extra fine	3230 (300)	2155 (200)
Machine operation, automatic	540 (50)	325 (30)
Meters	540 (50)	325 (30)
Missiles		
Repair and servicing	1075 (100)	540 (50)
Storage areas	215 (20)	110 (10)
General inspection	540 (50)	325 (30)
Office work, general	755 (70)	540 (50)
Ordinary seeing tasks	540 (50)	325 (30)
Panels		
Front	540 (50)	325 (30)
Rear	325 (30)	110 (10)
Passageways	215 (20)	110 (10)
Reading		
Large print	325 (30)	110 (10)
Newsprint	540 (50)	325 (30)
Handwritten reports, in pencil	755 (70)	540 (50)
Small type	755 (70)	540 (50)
Prolonged	755 (70)	540 (50)
Recording	755 (70)	540 (50)

*As measured at the task object or 760 mm (30 in.) above the floor.

Table 3.8. Specific Task Illumination Requirements (*Continued*)

Work area or type of task	Illumination levels, lux* (fc)	
	Recommended	Minimum
Repair work		
General	540 (50)	325 (30)
Instrument	2155 (200)	1075 (100)
Scales	540 (50)	325 (30)
Screw fastening	540 (50)	325 (30)
Service areas, general	215 (20)	110 (10)
Stairways	215 (20)	110 (10)
Storage		
Inactive or dead	55 (5)	30 (3)
General warehouse	110 (10)	55 (5)
Live, rough, or bulk	110 (10)	55 (5)
Live, medium	325 (30)	215 (20)
Live, fine	540 (50)	325 (30)
Switchboards	540 (50)	325 (30)
Tanks, containers	215 (20)	110 (10)
Testing		
Rough	540 (50)	325 (30)
Fine	1075 (100)	540 (50)
Extra fine	2155 (200)	1075 (100)
Transcribing and tabulation	1075 (100)	
	540 (50)	

*As measured at the task object or 760 mm (30 in.) above the floor.

NOTE:: Some unusual inspection tasks may require up to 10,000 lux (1000 fc). As a guide in determining illumination requirements, the use of a steel scale with 1/64-inch divisions requires 1950 lux (180 fc) of light for optimum visibility.

SOURCE: MIL-STD-1472D

8. Avoid stacking of components (the one that went bad is always on the bottom).
9. *Never* provide a screwdriver adjustment which is not clearly visible during adjustment.
10. All lubrication fittings and lubricant levels must be serviceable without disassembly.
11. Minimize the number of types of fasteners.
12. Provide captive fasteners where dropping or loss of the item may result in damage to equipment or danger to personnel.

Although it is not mentioned in the military specifications, designers should provide the capability for rendering the equipment inoperative whenever maintenance workers have to get any portions of their bodies

Table 3.9. Recommendations for Display Lighting

Condition of use	Lighting technique	Brightness of markings, cd/m^2 (Ft-L)†	Brightness adjustment
Indicator reading, dark adaptation necessary	Red flood, indirect or both, with operator choice	0.07–0.35 (0.02–0.1)	Continuous throughout range
Indicator reading, dark adaptation not necessary but desirable	Red or low-color-temperature white flood, indirect, or both, with operator choice	0.07–3.5 (0.02–1.0)	Continuous throughout range
Indicator reading, dark adaptation not necessary	White flood	3.5–7.0 (1–20)	Fixed or continuous
Panel monitoring, dark adaptation necessary	Red edge lighting, red or white flood, or both, with operator choice	0.07–3.5 (0.02–1.0)	Continuous throughout range
Panel monitoring, dark adaptation not necessary	White flood	35–70 (10–20)	Fixed or continuous
Possible exposure to bright flashes, restricted daylight	White flood	35–70 (10–20)	Fixed
Chart reading, dark adaptation necessary	Red or white flood with operator choice	0.35–3.50 (0.1–1.0)	Continuous throughout range
Chart reading, dark adaptation not necessary	White flood	17–70 (5–20)	Fixed or continuous

†Ft-L is the abbreviation of the unit foot-lamberts.

SOURCE: MIL-STD-1472D

into a position where moving parts of the equipment could cause injury if the equipment were started up accidentally. This is known as the *lock-out provision*. If parts of the equipment store mechanical, electric, or fluid energy, provision must be made to neutralize these energy sources during maintenance work. This feature is known as the *zero energy state*. It is intended to prevent maintenance workers from being injured by falling weights, loaded springs which suddenly snap back, electric capacitors which shock workers by discharging when touched, and sudden release of pressure from liquid or gas storage containers. Such design requirements are addressed by the National Safety Council and by OSHA.

One of the special features of MIL-STD-1472 is the inclusion of an index of "guidance documents." This listing is in addition to the mandatory callout given in the "Applicable Documents" section at the beginning of the standard. Use of the guidance documents is optional, and the references are intended only to provide assistance to the designer. They include six triservice publications, such as MIL-H-46855, and 17 Army publications, such as Army Regulation 385-16, *Safety for Systems, Associated Subsystems and Equipment* plus *Maintainability Guide for Design* (AD 823 539), listed as AMCP 706-134. There are 13 Navy publications entered, including Navships 94323, *Human Engineering Guidelines for Maintainability*. The Air Force has 32 entries, including seven design handbooks and *Guide to the Design of Mechanical Equipment for Maintainability* (AD 271 477, listed as ASD TR 61-381). Twelve voluntary standards, which are nongovernment publications, are listed including nine ANSI standards; two standards from the American Society of Heating, Refrigerating, and Air-Conditioning Engineers (ASHRAE); and the *Life Safety Code,* published by the National Fire Protection Association (NFPA) as its standard 101.

An extensive bibliography on the subject of human engineering lists 12 older standard reference works which are available from technical libraries. Those of particular value to this book include the following:

1. J. W. Garett and K. W. Kennedy, *A Collation of Anthropometry,* Aerospace Medical Research Laboratory, Wright-Patterson Air Force Base, Ohio, 1971 (two volumes, Library of Congress catalog card 74-607818)
2. *General Safety Requirements,* U.S. Army Engineer Manual 385-1-1, Department Documents, U.S. Government Printing Office, Washington, D.C., October 1987
3. H. P. Van Cott and R. G. Kinkade, eds., *Human Engineering Guide to Equipment Design,* Wiley, New York (LC catalog card 72-600054)
4. *Lighting Handbook,* Illumination Engineering Society (IES), 345 East 47th Street, New York, New York, Volume 1: Reference, 1984, Volume 2: Applications, 1987
5. Henry Dreyfuss, *Symbol Source Book,* McGraw-Hill, New York, 1972 (LC card 71-172261)
6. A. Damon, H. W. Stoudt, and R. A. McFarland, *The Human Body in Equipment Design,* Harvard University Press, Cambridge, Mass., 1966 (LC card 65-22067)
7. J. A. Roebuck, K. H. E. Kroemer, and W. G. Thomson, *Engineering Anthropometry Methods,* Wiley, New York, 1975 (LC card 74-34272)

Some more recent references on human factors include W. E. Woodson, *Human Factors Design Handbook,* McGraw-Hill, New York, 1981; E. J. McCormick, *Human Factors in Engineering and Design,* McGraw-Hill, New York, 1982; *Handbook of Human Factors,* Wiley, New York, 1987; and P. Tillman and B. Tillman, *Human Factors Essentials,* McGraw-Hill, New York, 1991.

Air Force Design Handbooks

As previously noted, the U.S. Air Force has prepared a set of seven design handbooks. *Human Factors Engineering,* DH 1-3, in two volumes covers the detailed requirements for the design of displays and controls, the integration of controls with displays, and maintainability. The designs are based on extensive investigations of human performance capabilities and limitations.

A set of design notes (DNs) gives the fundamental data for seeing, hearing, touching, and positioning, with some information on smelling and tasting. The ability of humans to acquire, process, and respond to information input is distilled into a set of guidelines for designers. The anthropometric information is very similar to that given in MIL-STD-1472 in the tables.

Vision

It has been estimated that 80 percent of human knowledge is acquired visually. The characteristics of the visual capabilities of an "average" person for design purposes are given in note 2B2 of *Human Factors Engineering,* USAF DH 1-3, as follows:

1. Ten colors, five sizes of figures, five different levels of light intensity, and two flicker rates can be distinguished from each other.
2. Type which is 0.083 inch high (6 point) can be read easily under 30 foot-lamberts of light.
3. The visual field extends about 130° vertically and 208° horizontally, with the maximum visual acuity at the center of the field.
4. It takes about 0.6 second to change focus from near to far.
5. About 30 minutes are required for the eye to adapt from daylight brightness to darkness.

6. Discomfort and impaired vision occur if glaring reflections or bright lights are located within 60° of the sightline.
7. Moving objects are harder to see accurately; an angular velocity of 50 degrees per second causes a loss of 43 percent of visual acuity.
8. Acceleration of the viewer decreases visual acuity, decreasing by 50 percent at 7*g*.

Hearing

Next to seeing, hearing is the most important sense used in information gathering by humans. Even though frequency and intensity are two physically measurable characteristics of sounds, the human perception of sounds is a subjective thing. Humanly perceived loudness (volume) and pitch (frequency) are not related to the physical quantities in a simple linear manner. Physiological studies have determined that the ordinary human ear can detect sounds which vary in frequency from about 20 cycles per second, or hertz (abbreviated as Hz), to about 20,000 Hz. The upper limit is known to be dependent on the age of the person, declining with advancing years.

The intensity of sounds is measured in decibels (abbreviated as dB). The basis for this scale is assumed to be the minimum perceptible sound, and it is given an intensity of 0.0002 microbar. This is a pressure of about 1/3,000,000 of a gram per square centimeter in the frequency range of greatest sensitivity. As sound intensity increases, a level is reached at which listening becomes uncomfortable. This takes place at about 120 dB. Further increases in energy level to 140 dB cause pain. Extensive tests over long periods indicate that sustained levels of 85 dB cause hearing impairment. A table of allowable durations of various sound levels is included in the OSHA regulations. During the development stages of a product, designers of equipment must consider these limitations on the amount of sound or noise generated by the product and take corrective actions if the limits are exceeded.

The adverse physiological effects of noise and excessive sound levels are well known. Sound may affect performance both positively—for pleasant sounds—and negatively—for sudden loud noises. Depending on what sort of task is being performed, sound may contribute to boredom, fatigue, or relaxation. Noise tends to increase muscular tension, making people edgy. This impedes performance on tasks requiring a high degree of muscular coordination or mental concentration. Continuous or rhythmically changing sounds of low frequency are less annoying than randomly or irregularly varying noises.

Where the design of a product requires that operators communicate with each other by means of speech, designers should know that a sound level of between 60 and 75 dB is usually used when the conversants are 36 to 40 inches apart. The frequencies used in human speech lie almost entirely between 100 and 8000 Hz, and over 50 percent of the sound energy is below 1000 Hz. It is common experience that the intelligibility of speech – meaning that the sound can be understood by the listener – decreases as the ambient noise level increases. Tests have indicated that the speech sound level must be at least 6 dB above the ambient noise level if a 75 percent intelligibility factor is to be realized.

Humans as Information Processors

The opening sentence in design note 2B8 of *Human Factors Engineering*, DH 1-3, neatly pigeonholes the place of humans in the system design concept. It states, "The role of the human being in the system is to detect information regarding some aspect of his environment, process this information in some manner, and transmit it in whole or in part to some response agency which affects appropriate adjustments in the environment." How is that for an ego trip? It appears to relegate people to the status of 160-pound black boxes.

Still, the designer of equipment must take into account the ability of operators to perform their required tasks after the tasks have been defined in the analysis stage of the design process. When the designer has determined what function the operator is to perform, then the types of information needed to perform the task can be determined and the method of presenting this information can be established. In most cases the visual or audible methods are used, since these two senses provide humans with most of their information intake capabilities. This fact, of course, is the basis for audiovisual educational methods and has been widely documented.

Guidelines for presenting information to operators include the following:

1. Present only essential information and avoid redundancy.

2. Use the most appropriate sensory channel.
3. Use known physical limitations to ensure that information is presented clearly.
4. Arrange visual displays to minimize the amount of eye movement required.
5. Use visual displays for:
   - Pictorial or tridimensional displays
   - Simultaneous comparisons
   - Small quantitative distinctions
   - Fast choices among many alternatives
   - Rapid scanning of information
   - Sending "difficult" material
   - Times when ambient noise level is high
6. Use audio information transfer for:
   - Fast two-way communication
   - Retention of familiar material
   - Caution, warning, or alerting signals
   - Interruption of an existing transmission
   - Times when vision is limited

After the information has been given to the recipient, that person has to evaluate it, make some sort of decision based on it, and determine what action may be necessary. This is the information processing task for humans. It places humans as a data link in human-machine systems, receiving information (input) from various channels and responding (output) to other parts of the system. The efficiency of the person can be measured by comparing the input with the output, where any differences are attributed to "noise" introduced by the person.

It is well known that the operational safety and the overall efficiency of myriad types of machinery, equipment, and products depend on the speed and accuracy of the users' or operators' responses to various information inputs. However, the speed and accuracy of human responses are heavily dependent on the quality and quantity of information given to the people and their ability to interpret it. Design note 2B9 of *Human Factors Engineering,* DH 1-3, gives some design guidelines for optimizing the information processing task and reducing the human "noise" contribution.

1. Reserve infrequently used channels for sending emergency information.
2. Identify the source of an audible message before it is given to the operator.
3. Use the hearing channel when considerable timesharing is involved.
4. Weight inputs according to their importance and probable frequency of occurrence, *and* tell the operator what the weighting scheme is.
5. Limit the number of variations of a given signal (pitch, color, intensity, etc.) to eight.

One of the outstanding features of *Human Factors Engineering,* DH 1-3, is the inclusion of an extensive bibliography of publications in human factors engineering. More than 200 references are given for materials available up to 1987. Most of the listed material was developed under research grants from a government agency, but some is from commercial sources or from the journals of learned societies.

Human Factors Engineering Programs

Not content with setting the human engineering requirements in MIL-H-46855 and the design criteria in MIL-STD-1472, the Department of Defense has prepared a supplement to the requirements, telling defense contractors how to comply with the requirements. *Human Engineering Procedures Guide,* MIL-HDBK-763, gives techniques and procedures for setting up a human factors program in an engineering department. Although it is directed to defense contractors, the information is sufficiently general to be of use to a wide variety of commercial engineering organizations as well.

Section 7 of the handbook provides a minutely detailed description of seven human engineering techniques. These include three items that all engineering departments should already be using: a design criteria checklist, engineering drawings, and models. Three other techniques have somewhat limited applications and include the development of visibility diagrams to show what areas the user can see from the operating position, a reach envelope to give a three-dimensional definition of the physical volume available to the operator, and the use of mannequins to simulate the presence of an operator.

For consideration in the early stages of the design effort, a listing of the comparative capabilities of humans versus machines is given. Twelve separate capabilities are described for which humans excel and twelve for which machines should be selected instead of humans. These are given in Table 3.10, which is also known as the Fitts list of human versus machine capabilities. Once the product designer has determined what tasks or actions are to be carried out by the user of the product, it should be possible to assign the various actions to either the human or the machine through use of the Fitts list. This will assist in defining the interface between the person and the product.

Human Factors Design Details

With the procedures for implementing a human factors engineering program given in HDBK 763, the requirements established by specification MIL-H-46855B, and the design criteria given by MIL-H-1472D, all that remains for the product designer to do is nail down the details. The AFSC design handbooks contribute significantly to this process for displays, controls, and maintainability. The Army, through its MIL-HDBK-759A, also treats the design details of consoles, displays, and controls, using the same basic principles as the Air Force, but adds details about knobs, switches, handwheels, and other mechanical aspects of such equipment.

One important consideration in the design of mechanical controls is known as the *direction of motion expectancy*. In other words, when a given control is actuated, what does the operator *expect* will happen as a result of having moved the control? Certain stereotypical reactions are expected:

- Clockwise rotation of a rotary control is expected to increase the output of the function.
- Forward, upward, or rightward movement of a lever is expected to increase the output of the function.
- On vehicles, speed is expected to increase when the accelerator is depressed and to decrease when the brake pedal is depressed.

Other conventional movement stereotypes are given in Table 3.11.

A small note of caution is in order regarding the motion expectancies. For all the entries it is assumed that the movement is a voluntary one on the part of the operator. Note, however, that sometimes controls are actuated involuntarily. This can occur if the operator moves the control accidentally, if a foreign object falls on the control, or if the operator fails to perform a motion called for by a predetermined program. When such mishaps occur, the equipment may malfunction in a manner

Table 3.10. Human versus Machine Capabilities

Humans Excel Machines in:
1. Detecting certain forms of energy at very low levels of intensity
2. Sensing an extremely wide variety of signals
3. Perceiving and analyzing patterns and drawing conclusions from them
4. Storing large amounts of information for long periods and recalling relevant facts when required
5. Exercising judgment and making decisions when events cannot be completely defined
6. Improvising when conditions change beyond usual limits and adopting flexible procedures
7. Reacting correctly to unexpected low-probability events
8. Developing original alternative solutions to problems
9. Utilizing information feedback to learn from experience and alter subsequent course of action
10. Performing fine manipulations, especially where misalignments appear unexpectedly
11. Continuing to perform when overloaded
12. Reasoning inductively
Machines Excel Humans in:
1. Monitoring performance in both humans and machines
2. Sensing signals beyond the range of human responses (infrared, ultrasonics, radio waves)
3. Performing routine, repetitive, or very precise operations with high accuracy
4. Storing and recalling large amounts of information in short time periods
5. Operating in hostile environments which are beyond human-tolerance limits
6. Repeating operations identically, precisely, continuously, and rapidly for long periods
7. Responding very rapidly to control signals
8. Carrying out several tasks simultaneously
9. Insensitivity to extraneous factors or distractions
10. Exerting large amounts of force smoothly and precisely with excellent repeatability
11. Performing complex computations very rapidly and with high accuracy
12. Processing information deductively

Table 3.11. Control Movement Stereotypes

Direction of Movement	Function
Up, right, forward, clockwise, pull	On
Down, left, rearward, counterclockwise, push	Off
Clockwise, right	Right
Counterclockwise, left	Left
Up, back	Raise
Down, forward	Lower
Up, rearward, pull	Retract
Down, forward, push	Extend
Forward, up, right, clockwise	Increase
Rearward, down, left, counterclockwise	Decrease
Clockwise (valves)	Close
Counterclockwise (valves)	Open

which is hazardous to the operator or to the equipment or to both. The designer must recognize that such events can take place and design the resulting hazards out of the equipment.

An example of such design is a foot-operated electric switch. Switches of this type are actuated by the operator pressing down on a lever. Movement of the lever closes the contacts in the switch, allowing power to flow through the switch. If the operator accidentally steps on the switch or if a heavy object happens to fall on it, the switch will be actuated and the equipment will operate when it may be expected not to do so. To prevent such a surprise, such switches are designed with a heavy protective cover over the top and down the sides. The cover protects against the falling-object hazard and forces the operator to make a conscious choice to insert his or her foot under the cover to reach the lever. Some models also include a spring-loaded kick plate at the rear of the cover which forces the operator to make an additional conscious effort to push her or his foot into the operating position.

Examples of programmed controls are found in the design of "dead man" switches which come into play when the operator becomes inattentive, falls asleep, drops dead, or is incapacitated in some manner. The programmed control then takes over, causing the equipment to assume a benign condition.

Whereas the Air Force design handbooks (DH 1-3) treat the information processing function of humans in somewhat general terms, the

Army treats them in great detail in MIL-HDBK-759A. More than 170 pages are devoted to the design of controls, displays, and consoles. Charts and tables show preferred designs, with information on the proper spacing between controls. Twenty-four illustrations and eighteen tables give detailed data on handwheels, key locking devices, a dozen styles of knobs, levers, joysticks, toggle and rocker switches, keyboards, and dual-axis controllers. Several styles of displays are described in detail, including scalar indicators, cathode ray tubes, large-screen displays, and displays utilizing speech and audible signal inputs. The guidelines for audible signals are similar to those in the AFSC documents, but HDBK-759A treats them in more elaborate detail. An example of the fineness of the treatment is given in the section relating to the use of deemphasis in earphones and loudspeakers. Paragraph 1.2.7.3.2.4 states:

> *Use of De-emphasis* If transmission equipment employs pre-emphasis and peak-clipping is not used, reception equipment should employ frequency de-emphasis of characteristics complementary to those of pre-emphasis only if it improves intelligibility, i.e., de-emphasis should be a negative slope frequency response not greater than 9 dB per octave over the frequency range 140 to 4,800 Hz.

Consoles are also treated in detail with several tables and illustrations of preferred dimensions. Both standing and sitting operator positions are specified, with design guidelines given for both types. The layout and design should include

- Contouring and sloping the console or instrument panel to minimize parallax errors
- Locating controls for easy manipulation
- Providing adequate space and support for the operator at the console

Displays on consoles should be designed in accordance with the following principles:

- Place frequently monitored displays within 15° left or right of the centerline and between 0° and 30° below the operator's eye level. This is known as the *preferred position.*
- Use the preferred position for indicators which are used for long, uninterrupted periods.
- Viewing distance to displays should be not less than 16 inches and preferably at least 24 inches.
- Mount displays perpendicular to the line of sight to reduce parallax errors.

Guidelines for controls which are mounted on the consoles are based on the following principles:

- Locate primary controls between shoulder level and waist height.
- Place frequently operated controls at right front and left front of the operator.
- Group frequently used controls, locating them for right-handed operation and within 16 inches of the normal work position.
- Positions of controls should be verifiable visually, regardless of the operator's viewing angle.
- When the operator must manipulate a control while monitoring a display, locate the control close to and directly under that display.

Workspace Dimensions

In the beginning of this chapter we mentioned that the first objective of the controlling design standard MIL-STD-1472 was that "The product shall provide a suitable work environment." One of the basic requirements for a suitable work environment is that the space provided for workers be large enough for them to perform the required tasks. This obviously requires that the size, weight, and motion capabilities of human operators be known.

The sizes and weights of humans have already been given in tabulated anthropometric data from both the SAE and military sources. Additional data of a similar nature are given in HDBK-759A, but need not be repeated. The additional information on motion capabilities also comes from HDBK-759A. It is given in section 2.3. As nearly as can be determined, the data were obtained from a sample of male adults. The data are given in Table 3.12 and Fig. 3.10.

Human Strength Capabilities

While the layout dimensions of controls, displays, and consoles depend on corresponding human dimensions, anthropometric data are not sufficient, by themselves, to permit the designer to complete her or his calculations. The ability of the operator to move the controls must also be considered. The general design criterion is that 95 percent of the expected population should be able to operate the controls. The corollary is that the maximum required operating force should be exertable by the weakest 5 percent of that population. Data for hand, arm, and fin-

Table 3.12. Range of Human Motion

Body member	Movement	Lower limit, degrees	Average, degrees	Upper limit, degrees
A. Wrist	1. Flexion	78	90	102
	2. Extension	86	99	112
	3. Adduction	18	36	27
	4. Abduction	40	54	47
B. Forearm	1. Supination	91	113	135
	2. Pronation	53	77	101
C. Elbow	1. Flexion	132	142	152
D. Shoulder	1. Lateral rotation	21	34	47
	2. Medial rotation	75	97	119
	3. Extension	47	61	75
	4. Flexion	176	188	190
	5. Adduction	39	48	57
	6. Abduction	117	134	151
E. Hip	1. Flexion	100	113	126
	2. Adduction	19	31	43
	3. Abduction	41	53	65
	4. Medial rotation (prone)	29	39	49
	5. Lateral rotation (prone)	24	34	44
	6. Lateral rotation (sitting)	21	30	39
	7. Medial rotation (sitting)	22	31	40
F. Knee flexion	1. Prone	115	125	135
	2. Standing	100	113	126
	3. Kneeling	150	159	168
G. Foot rotation	1. Medial	23	35	47
	2. Lateral	31	43	55
H. Ankle	1. Extension	26	38	50
	2. Flexion	28	35	42
	3. Adduction	15	33	24
	4. Abduction	16	30	23
I. Grip angle		95	102	109
J. Neck flexion	1. Dorsal (back)	44	61	88
	2. Ventral (forward)	48	60	72
	3. Right	34	41	48
	4. Left	34	41	48
K. Neck rotation	1. Right	65	79	93
	2. Left	65	79	93

*These values are based on the nude body. The ranges are larger than they would be for clothed personnel.

Flexion: Bending, or decreasing the angle between parts of the body, Extension: Straightening, or increasing the angle between parts of the body, Adduction: Moving toward the midline of the body, Abduction: Moving away from the midline of the body, Medial rotation: Turning toward the midplane of the body, Lateral rotation: Turning away from the midplane of the body, Pronation: Rotation the palm of the hand downward, Supination: Rotation the palm of the hand upward

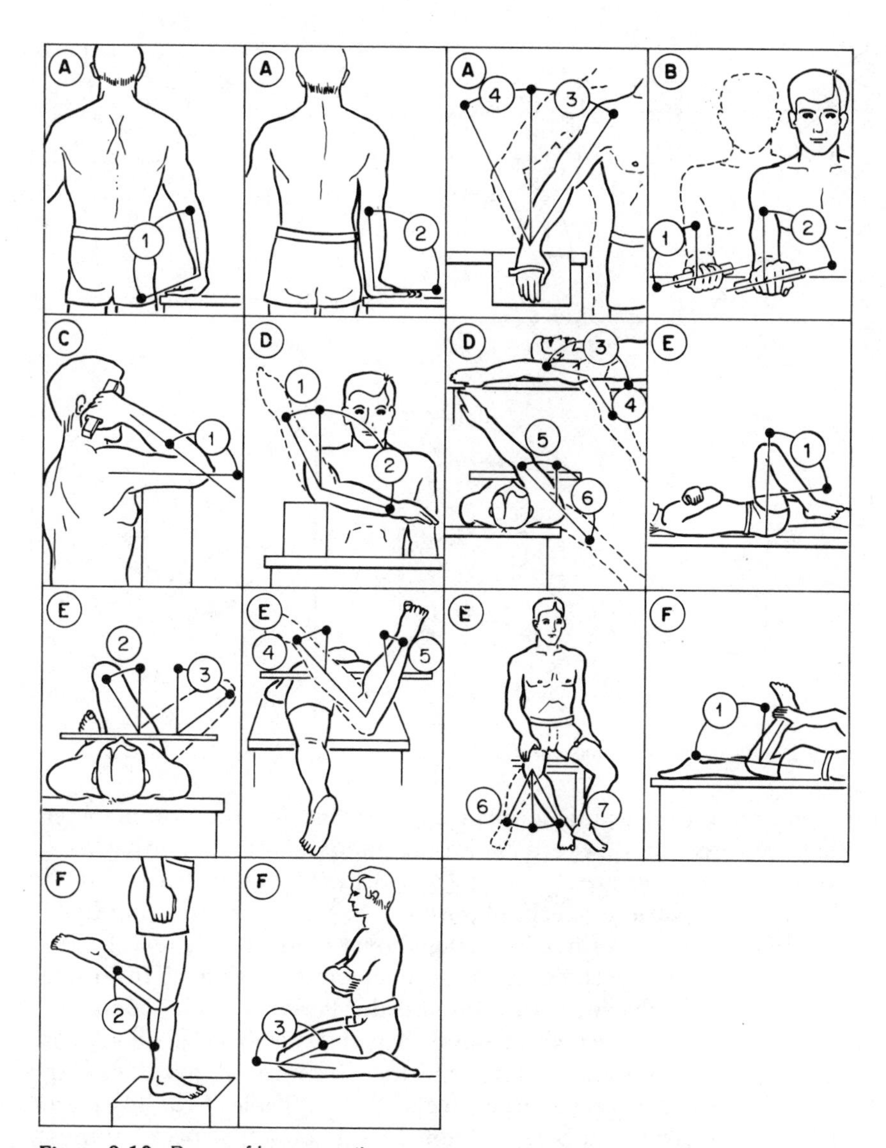

Figure 3.10. Range of human motion.

ger strength for the 5th percentile male are shown in Fig. 3.11. Leg strength at various angles is shown in Fig. 3.12. The values should be reduced by approximately one-third to apply to women. Push and pull forces can be exerted either horizontally or vertically. Horizontal forces may be applied momentarily (to move an object or set it in motion) or in a sustained fashion. The force limits given in Table 3.13 are based on

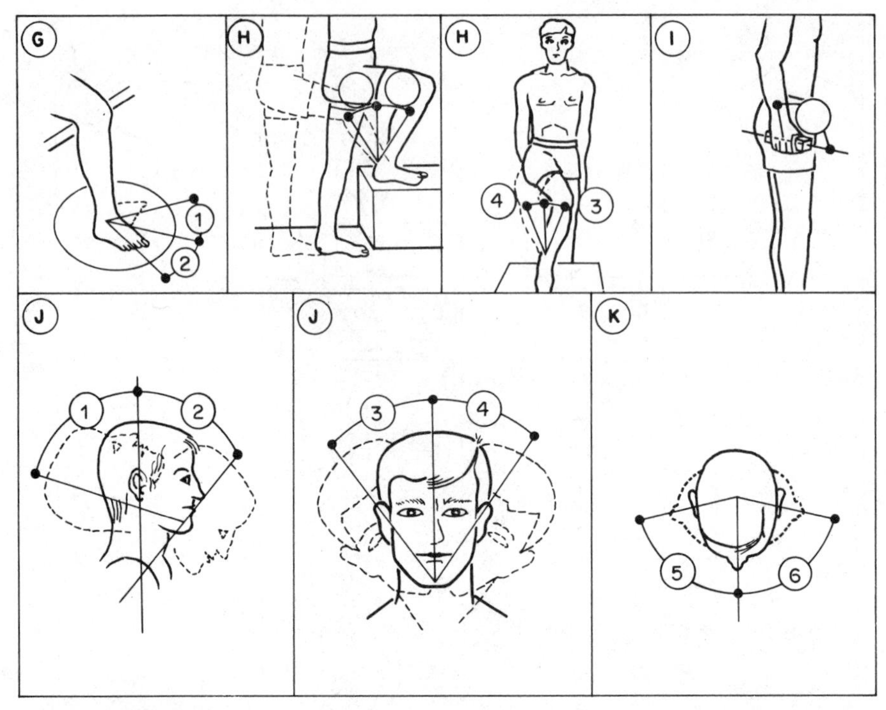

Figure 3.10. Range of human motion. (*Continued*)

the assumption that a suitable surface is available as a reaction member so that the force can be exerted with the hands, shoulders, or back.

Vertical force limits are shown in Table 3.14. The values are based on static pulls from various specified positions exerted against a ring fixed to the floor. The mean force is averaged over a 3-second interval. Lifting, however, is not a static activity but involves motion of the force through a certain distance. In addition, the location of the center of gravity of the lifted object with respect to its distance from the body has a dramatic effect on human lifting capacity. This is shown graphically in Fig. 3.13. Maximum weight limits are shown in Table 3.15. They are very important. National Safety Council data show that back injuries, caused by improper lifting, are the largest single cause of industrial lost-time injuries. They are preventable if weight limits are observed and proper lifting technique is used.

Safety as a Human Factor

As mentioned previously, MIL-STD-1472 deals only superficially with the concept of safety as a separate item. It relies on industrial, commer-

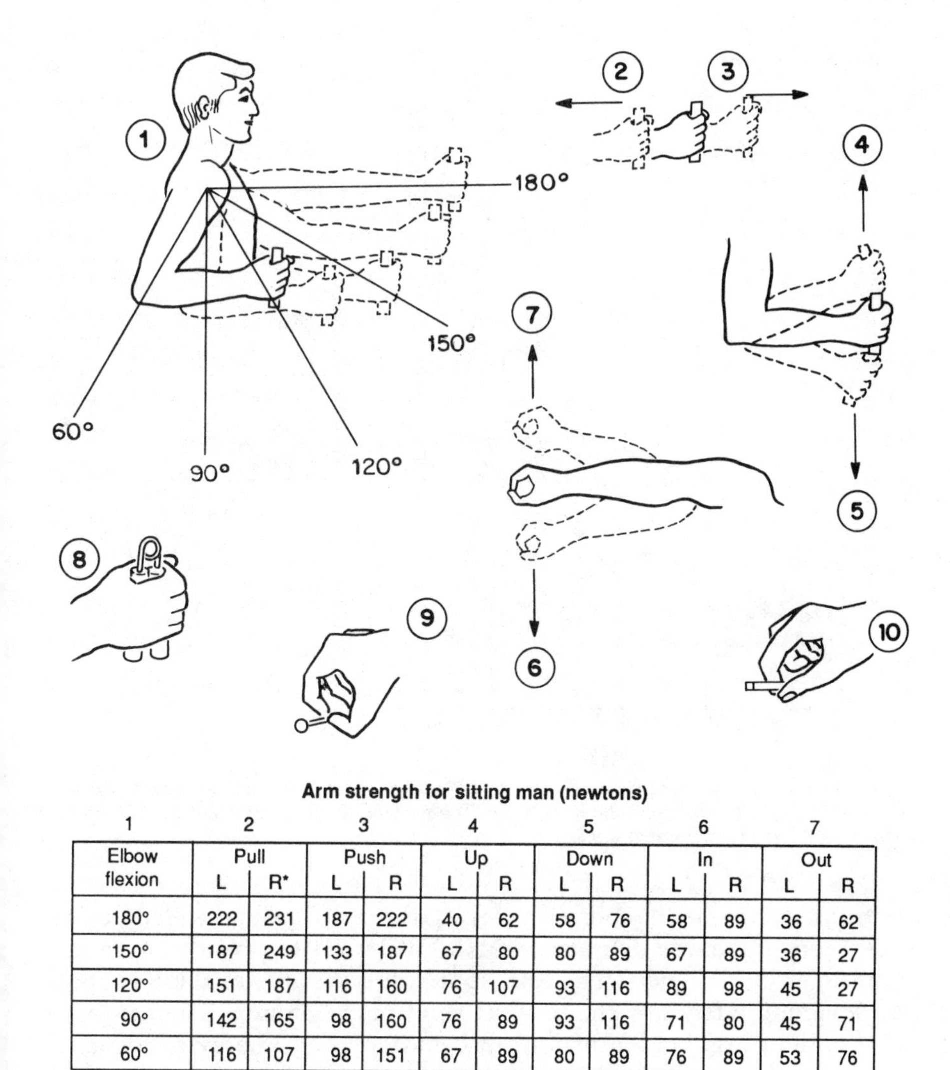

Arm strength for sitting man (newtons)

1	2		3		4		5		6		7	
Elbow flexion	Pull		Push		Up		Down		In		Out	
	L	R*	L	R	L	R	L	R	L	R	L	R
180°	222	231	187	222	40	62	58	76	58	89	36	62
150°	187	249	133	187	67	80	80	89	67	89	36	27
120°	151	187	116	160	76	107	93	116	89	98	45	27
90°	142	165	98	160	76	89	93	116	71	80	45	71
60°	116	107	98	151	67	89	80	89	76	89	53	76

Arm strength for sitting man (newtons)

	8		9	10
Holding time	Hand grip		Thumb-finger grip (palmar)	Thumb-finger grip (tips)
	right	left		
Momentary hold	250	200	60	60
Sustained hold	145	155	35	35

*L = left; R = right

Figure 3.11. Arm, hand, and thumb-finger strength (5th percentile male).

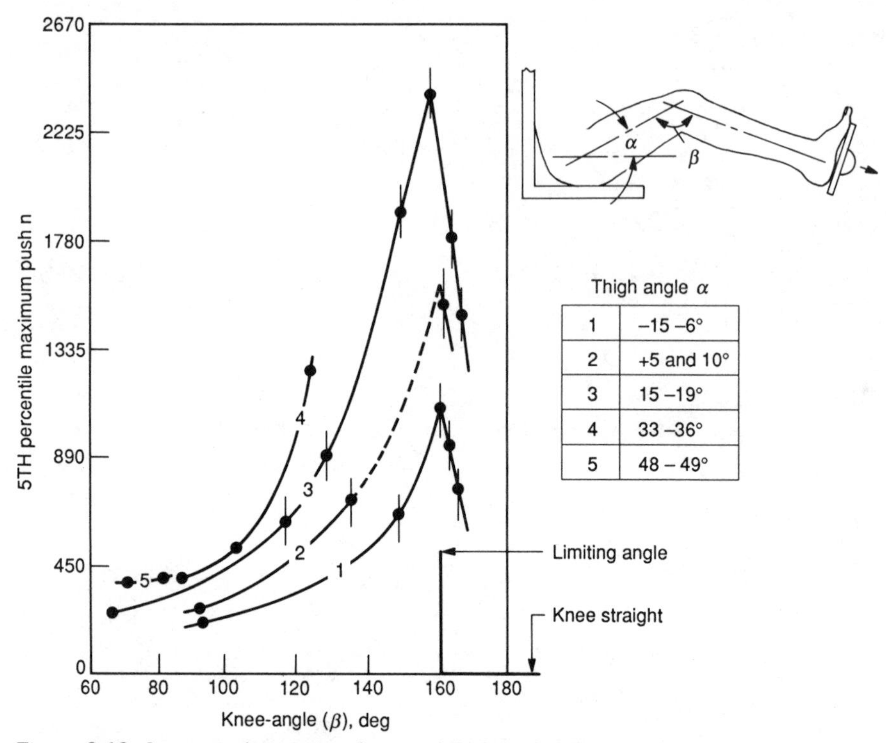

Figure 3.12. Leg strength at various knee and thigh angles. The push force exerted by the leg depends on the thigh angle and the knee angle. The maximum push is at about the 160° knee angle, referred to as the ***limiting angle.***

cial, and government publications for its sources of recommended practices and standards. However, MIL-HDBK-759A provides several guidelines for designers. The basic approach is the long-established one of designing safety into the product from the beginning. The new aspect is the consideration of human behavior as an important human factor for the designer.

Because most people believe that accidents always happen to somebody else, carelessness is a common human factor which the designer must consider in the concept stages. Many extensive investigations of accident causation have been carried out by both military and civilian organizations to affirm the significance of carelessness as a major contributing factor to personal injuries or damage to equipment. These investigations have shown some of the reasons why people make mistakes, misuse or abuse equipment, or engage in unsafe practices. As guidelines for designing goofproof products, the following principles have been developed:

Table 3.13. Horizontal Push and Pull Forces Exertable (Intermittently or for Short Periods)

Horizontal force,* at least	Applied with:	Condition (μ is coefficient of friction)
110 = N push or pull	Both hands or one shoulder or the back	With low traction $0.2 < \mu < 0.3$
200 = N push or pull	Both hands or one shoulder or the back	With medium traction $\mu \sim 0.6$
240 = N push	One hand	If braced against a vertical wall 510 to 1520 mm from and parallel to the push panel
310 = N push or pull	Both hands or one shoulder or the back	With high traction $\mu > 0.9$
490 = N push or pull	Both hands or one shoulder or the back	If braced against a vertical wall 510 to 1780 mm from and parallel to the panel *or* If anchoring the feet on a perfectly nonslip ground (like a footrest)
730 = N push	The back	If braced against a vertical wall 580 to 1090 mm from and parallel to the push panel *or* If anchoring the feet on a perfectly nonslip ground (like a footrest)

*May be doubled for two (and tripled for three) operators pushing simultaneously. For the fourth and each additional operator, not more than 75 percent of her or his push capability should be added.

1. Recognize and identify actual or potential hazards, then design them out of the product.
2. Thoroughly test and evaluate prototypes of the product to reveal any hazard missed in the preliminary design stages.
3. Make certain that the product will actually perform its intended function in an acceptable manner so that the user will not be tempted to modify it or need to improvise possibly unsafe methods for using it.
4. If field experience reveals a safety problem, determine its real cause, develop a corrective action to eliminate the hazard, and follow up to make certain that the corrective action is successful.
5. Design equipment so that it is easier to use safely than unsafely.
6. Realize that most product safety problems arise from improper product use rather than product defects.

Table 3.14. Static Muscle Strength Data

Strength measurements	Percentile values, newtons			
	5th percentile		95th percentile	
	Women	Men	Women	Men
A. Standing two-handed pull: 38 cm level				
Mean force	330.9	737.5	817.6	1354.5
Peak force	396.9	844.7	888.3	1437.2
B. Standing two-handed pull: 50 cm level				
Mean force	326.1	758.0	840.7	1341.6
Peak force	374.1	830.9	905.2	1441.7
C. Standing two-handed pull: 100 cm level				
Mean force	185.0	444.4	443.0	931.0
Peak force	218.0	504.0	493.3	988.4
D. Standing two-handed push: 150 cm level				
Mean force	153.5	408.8	379.9	1016.9
Peak force	187.7	472.8	430.1	1094.3
E. Standing one-handed pull: 100 cm level				
Mean force	102.8	214.8	283.8	627.6
Peak force	131.7	258.9	322.5	724.2
F. Seated one-handed pull: centerline, 45 cm level				
Mean force	106.3	222.3	391.9	678.4
Peak force	127.2	273.1	450.6	758.4
G. Seated one-handed pull: side, 45 cm level				
Mean force	109.0	239.8	337.2	603.6
Peak force	134.3	272.7	394.6	658.8
H. Seated two-handed pull: centerline, 38 cm level				
Mean force	241.5	594.7	770.4	1221.0
Peak force	284.7	698.8	841.6	1324.2
I. Seated two-handed pull: centerline, 50 cm level				
Mean force	204.2	524.9	631.6	1052.0
Peak force	237.1	595.6	687.0	1189.0

Several organizations have been collecting accident reports for decades and have recognized certain recurrent types of incidents and causes. This effort has permitted the preparation of checklists to assist designers in recognizing possible hazards early in their programs. The typical questions which designers should ask themselves are listed in Table 3.16.

Summary

The part of the equation which balances the cleverness, creativity, ingenuity, and skill of product designers against the wants, needs, and capabilities of the ultimate user is the human factor. No matter how they may try to avoid it, designers must ultimately accept the human being as

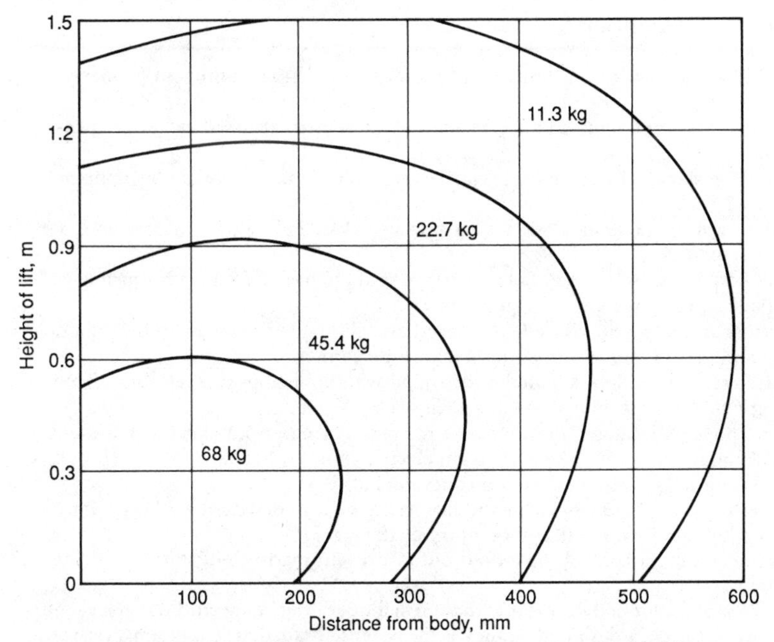

Figure 3.13. Manual lifting capacity: lifting forces that can be exerted by 95 percent of personnel (male), using both hands.

Table 3.15. Maximum Weight Limits

Type of handling	Height lifted, meters				
	0.3	0.6	0.9	1.2	1.5
Lifting,* kg					
One person	38.6	36.3	29.5	22.7	15.9
Two persons	77.1	72.6	59.0	45.4	31.8
Carrying (five steps or less),† kg					
One person	29.5				
Two persons	59.0				

*These weight limits should be used as maximum values in establishing the weights of items that must be lifted. These limits apply to items up to 380 mm long and up to 305 mm high, with handles or grasp areas. These limits should not be used for larger items, or for items which must be lifted repetitively.

†These weight limits should not be used if personnel must carry the item more than five steps.

SOURCE: MIL-HDBK-759A, June 30, 1981.

Table 3.16. Design Safety Checklist

1. Are all moving parts of machinery and power transmission equipment properly guarded?
2. Are edges of components and access openings rounded, or are sharp edges protected?
3. Are audible signals distinctly recognizable and unlikely to be masked by ambient noise?
4. Are fault-warning systems designed to detect weak or failing parts before an emergency occurs?
5. Are all liquid, gas, and steam lines clearly identified and labeled, with warnings of specific hazards to persons or equipment?
6. Are struts and latches provided for hinged and sliding components to keep them from shifting and endangering maintenance personnel?
7. Are drawers and foldout assemblies provided with limit stops to keep them from coming out too far, coming loose, or falling?
8. Are components located and mounted so maintenance personnel can get at them easily without exposure to hazards from electric charges, hot surfaces, sharp points or edges, moving parts, or chemical contamination?
9. Are mechanical components which utilize heavy springs designed so the springs cannot suddenly fly out and inflict injury or damage?
10. Can the equipment controls be locked out to prevent inadvertent start-up during maintenance?
11. Are indicators color-coded to show the normal operating range and danger range?
12. Are warning circuits designed so they actively sense hazardous conditions, rather than passively display control settings?
13. Are the most critical warning indicators grouped within the operator's normal field of view and separated from other, less important indicators?
14. Are go/no go or fail-safe circuits used wherever possible to ensure that failures do not produce additional hazards and that operators know that a failure has taken place?
15. Are charge bleed-off devices provided for high-voltage capacitors which may be touched by personnel during servicing?
16. Are adjustment screws and commonly replaced parts located away from high voltages or hot surfaces?
17. Are conspicuous warnings placed near high-voltage or high-temperature components?
18. Are safety interlocks used where necessary?

the independent variable over which they have absolutely no control. The person who buys, uses, or operates the result of designers' efforts simply comes as he or she is—warts and all.

Fortunately, an abundance of information is available to designers regarding the characteristics of a wide spectrum of people. There are excellent sources of detailed information on human dimensions, sensory abilities, and physical attributes. Some limited data are available regarding mental and psychological characteristics that may be of use to designers.

A large body of experience has been accumulated regarding accident causation. This information has been classified according to frequency of occurrence and severity of loss for a time span of several decades. The results have been applied to the development of design standards and the promulgation of regulations, with the objective of protecting persons and property from injury. The influence of human factors in the formulation of these publications has been profound, although it often is latent rather than apparent.

There are well-established methods and procedures available to assist designers in analyzing the functions of any product, determining the hazards present in any specific task, and formulating ways of designing the hazards out of the product in the beginning stages of the design. With the present state of knowledge regarding human factors, there is absolutely no excuse for any designers to neglect inclusion of this independent variable in the generation of solutions to their design problems.

4
The Product Design Process

The products considered in this chapter are limited to those involving primarily mechanical, electrical, and structural considerations. Chemical, biological, and nuclear products do not generally fall within the categories of products considered here, even though all may involve serious safety problems. These safety problems are addressed in many government regulations and industry publications which are outside of the scope of this book.

The techniques available for the design of products have advanced enormously in the latter portions of the 20th century. The introduction of fast, reliable computational machines which began in about 1950 has increased the sophistication and power of analytical methods for the solution of mechanical and electrical problems. The slide rule—once the symbol of the engineer—is now a relic, and the mechanical calculators, ingenious though they were, are museum pieces. With present computational methods it is no longer acceptable to come up with just one possible solution to a design problem. The power of the computer permits many possible solutions to be generated. This multiplicity of choice provides the designer with previously unavailable insight into the nature of the problem, thus permitting the selection of the optimum solution from the available list.

Problem Statement

With the ability to analyze problems rapidly and to develop many possible solutions comes the necessity for clear and unambiguous formulation of the statement of the problem to be solved. Without rigorous definition of what—exactly—is wanted, the optimization process may not work properly. Proper optimization calls for the highest levels of the designer's skills in order to take advantage of the insight provided from the computational results. Without clear criteria delineated in the problem statement, optimization may degenerate into firing at random and hoping to hit it.

While a few products may be classified as solutions looking for a problem, the vast majority of products are conceived in response to some real or perceived need. The need is usually recognized by the people responsible for marketing the company's products. Need may be seen in a customer seeking a product not presently available, dealers wishing to expand their product lines, field reports of a new item introduced by a competitor, or new product development efforts in the company's R&D department. Sometimes the need arises from requirements to reduce costs, or enhance value, of a given product. Other needs may be the result of obsolescence of the existing product and the necessity for updating it with a new design. Whatever the source of the real or perceived need, that need must be defined as precisely as possible. Any assumptions made in defining the need should be spelled out and put on record, since no product design is going to be better than the underlying assumptions on which it was based.

Definition of the need will answer questions involving who will use the product, the manner in which it will be used, and any special characteristics the product should have. This information automatically sets the objectives as to what the product is supposed to do and how it is supposed to do it. From that base, the statement of the design problem can be drawn up.

While the definition of the need is usually the responsibility of the marketing department, formulation of the problem statement should involve input from all major departments of a company. The problem statement is usually formulated by the engineering department, since people in that department will be working to generate a suitable solution to the stated problem. However, the marketing people should be aware of the impact of this new product on the existing product line; the legal department should be made aware of possible product liability problems and opportunities for patent protection or licensing agreements; the finance department must be made aware of money requirements and their timing, the manufacturing department should have information on the materials and processes which are likely to be required. All these departments should be able to contribute valuable perspectives to the problem statement, reflecting their own concerns.

Some concerns external to the company may be relevant to the problem statement, particularly if the product has any impact on the environment. Often consideration must be given to the expected useful life of the product and how it will be disposed of when the end of that useful life has been reached. Heavy metals and toxic chemical components pose particular concerns since their disposal presents potential severe adverse environmental effects. Recyclability of materials is receiving more and more attention as disposal sites are filling up, with the result that life-cycle costs of the product are finding their way into product problem statements.

Proposed Solutions

Once the statement of the design problem has been formulated, reflecting all the inputs from various company departments, the actual work of developing solutions to the problem can begin. (Note that the word *solutions* is used in the plural, not singular.) If the proposed product is a redesign or update of an existing product, there is an ample base from which to work, by using the problem definition for direction and the existing product as the point of departure. The amount of creative thinking and solution synthesis required will be relatively small in contrast to the effort required when a really "new" product is under consideration.

For a product which does not have an existing reference base, the design process begins with the determination of the motions required to produce the desired result. Once the motions have been defined, the linkages to produce those motions can be synthesized. This is the *kinematics* part of the design, and it may involve several possible solutions. When the designer has chosen what is considered the "best" kinematic solution, the proposed mechanism can be drawn up as a set of lines which are connected in a suitable manner. The next step is to analyze the velocities and accelerations of the various components, based on the selection of a certain operating speed for the device. This is the dynamic part of the analysis, and it permits the determination of the inertia forces generated by the motion of the parts themselves.

The next step is to determine the forces which will be applied to the components during the operating cycle, while the various parts are doing their intended work. With the applied forces and the dynamic forces known, the various parts can be proportioned to withstand the known forces. This is known as the *strength-of-materials* portion of the design process, and it results in the set of lines from the kinematics solution being given cross sections and mass from which to calculate the inertia forces mentioned previously.

In performing the strength-of-materials analysis, the designer is forced to consider which specific material is to be selected for the particular application. The mechanical properties of this selected material must be known so that suitable values for the working stresses can be assigned. The working stresses must reflect the effects of repeated load applications—the fatigue phenomenon-and the effects of changes in cross section, where reentrant corners may be present which generate large stress concentrations. Several reference books are available which provide tables and charts for the solution of many stress analysis problems, including stress concentrations factors and the effects of fatigue. In all stress analysis work, close attention must be paid to the selection of appropriate factors of safety. Some designers refer to safety factors as "factors of ignorance," reflecting how little is actually known about what goes on during the life of a machine component. In many instances, the factors of safety to be used are specified in codes or standards which apply to the particular product.

In some design problems, the effect of deflections of the parts is important, and their elastic properties must be taken into consideration. Elaborate methods have been devised for the determination of deflections under applied loads and are given in the standard texts on the subject. Deflection analysis is the key to the analysis and design of statically indeterminate structures, and it has been incorporated into many elaborate computer programs for the design of such structures.

For many applications the weight of the parts is of primary importance. Aircraft products are the obvious example because of the adverse effects of weight on the payload of the aircraft. However, ships also require extensive and intricate weight calculations to ensure the proper trim and to permit the calculation of the metacentric height for stability in the rolling mode. A high level of designer skill is required when weight problems arise, particularly when weight must be taken out of a product. Knowing where and how much to shift requires real insight and knowledge of the laws of mechanics.

When the design calculations have been completed, the usual procedure is to make a sample of the proposed product, based on preliminary design drawings which have been checked but not released. The sample presents a three-dimensional model of the proposed product for examination and review. The sample may actually be nothing more than a good mock-up if the product is very complex, or it may be a prototype which can be used as a "proof of principle" device to establish the feasibility of the proposed design. Such a device may be used to answer the question, will it work? If the answer is yes, the design is ready for its initial design review.

Design Review 1

Just as the formulation of the problem statement involved inputs from all major departments of the organization, so, too, the initial design review requires a similar information exchange. Since the engineering department has made the greatest effort, the review session should normally be called and directed by this group. All the pertinent factors should be discussed at the review sessions, including estimated costs, manufacturing methods and equipment to be required, and an evaluation of how well the proposed product meets the criteria set forth in the problem statement. In addition, during the first design review any hazards present in the proposed product should be recognized. Hazard recognition at the earliest possible stage is a key element in the procedure for designing a reasonably safe product.

Hazard recognition requires considerable background and training in the field of accident causation. Unfortunately, there is no formal academic training in this specialized area, but the National Safety Council (NSC) and other organizations publish information on the subject. In addition, many safety standards have been published by the American National Standards Institute (ANSI), which can be used as reference materials, and there are magazines and periodicals devoted to the subject of safety which have articles of interest to the safety professional.

A key element in the hazard recognition process is the question of foreseeability. The design engineer, working from the problem statement and attempting to meet its requirements, naturally assumes that the product is going to be used in a certain manner by the ultimate purchaser. The engineer naturally wants the user to employ the product in the way assumed by the designer so that the product will perform up to its best capabilities. *But,* and this is a big but, the user will usually not know anything about the engineer's assumptions and will use the product in her or his own fashion. It is one of the jobs of the people conducting the design review to question the assumptions of the design engineer and to attempt to foresee other possible modes of use for the proposed product.

The foreseeability exercise is really an effort to apply what is euphemistically called "Murphy's law" (i.e., if something can go wrong, it will). Ideally, the hazard recognition effort will be directed by a single individual who has the responsibility for the safety aspects of the proposed product. This person will act as a professional klutz and will use, misuse, and abuse the product in every way that can be foreseen. If assembly is required, the tester will put the thing together backward if it is possible to do so. If there is an exposed hot part, the tester will burn himself or herself on it. The tester will get an electric shock from any exposed electric wiring and will get cut on any sharp edges. In addition, this person will try to do things with the product that the problem statement and the design engineer never intended. Any dangers to the operator which are revealed during this process fall into the class of recognized hazards. Their existence calls for some sort of corrective action on the part of the designer.

The best type of corrective action is to redesign the product to eliminate the recognized hazard. This process is called designing out the hazard. If the hazard cannot be designed out of the product, the next best approach is to design suitable guards to protect the operator from the hazard. This is the "guarding out" process. If the hazard cannot be designed out or prevented by guards, then a third method is used. This is the least desirable approach, and it consists of warning out the hazard by providing suitable labels, decals, signs, or markings to make the operator aware of the existence of the hazard.

As soon as the hazards are recognized and identified, they should be classified as to their severity. This information should then be given to the technical writers who will be preparing the operator's manuals and servicing instructions. The writers can then include suitable notices in their text material to alert the operator or service person to the existence and nature of the hazard.

Note, please, that all this safety effort is intended to be carried out in the preliminary design stages of the product development cycle, not as an add-on after the final design has been frozen. It is far more cost-effective to design in the safety features at the beginning of the design effort than to make "fixes" later on.

Detail Design

Once the first design review has been completed and agreement has been reached to continue the project, the required changes can be cranked into the design process. These changes will include the safety features previously mentioned and all necessary cost, producibility, and serviceability requirements generated by the design review process. The details of the design can now be specified, including the selection of the materials, the processing steps which will be required, and the determination of fits, forms, and finishes of each component. It is during this phase of the design effort that the personal knowledge and skill of the designer comes into play most vigorously. Selection of vendors for purchased items begins as soon as enough design details become available to permit vendors to interpret the drawings in a meaningful manner.

The vendors will often have comments on the design, and there will usually be extensive communication between the designers and the vendors to iron out the details.

Control of the design process for a particular product is often achieved through bar charts of the Gantt type or through Program Evaluation and Review Technique (PERT) charts. These documents indicate the status of the design effort at any time, and they can provide the supervisors with information regarding which, if any, parts of the program need more or less attention. These methods of program management are well described in the available literature and are not discussed further here.

Once a complete set of parts is available, construction or assembly of the "first article" can be carried out. This is often a major event in the program, since the efforts of the designers have finally come to fruition in one place and the designers can see the results of their contributions. This should be a source of satisfaction to all concerned.

Now comes the testing and evaluation phase of the program. How well does the product meet the requirements of the problem statement? How close to the estimate did the cost of the product come? And how long will the product last in service? All these and many other questions must be answered by the testing and evaluation program. In most cases there will be component failures during the testing program, sometimes many of them, but this is to be expected. The absence of any component failure is considered by some engineers as a certain indication that the product is overdesigned, and refinements are necessary. Redesign, or "back to the drawing board," is the natural result of a good testing program and should not be resented by the designer. Obviously, it is far better to find and eliminate defects in the product at the testing stage than to let the customer find them later.

There are many commercial products for which the published standards include a detailed requirement for the testing of the product. Portable ladders and office furniture are two common examples in which the standards call out detailed testing procedures. Of course, the existence of these standards and the testing requirements should have been recognized in the problem statement at the beginning of the project, so the designer should not be surprised by them during the testing and evaluation program. Almost all government and military products have detailed testing requirements included in their specifications. A definition of what constitutes a failure is often incorporated in the text of the specification.

While performance, endurance, and reliability are each an important factor to be evaluated during the testing, the hazards to the user must also be investigated at that time. Hazard recognition, which was emphasized during the first design review, should be an inherent part of the evaluation program. The person responsible for the safety aspects of the product should design and carry out an "idiot-proofing" program during the testing effort to make certain that the hazards recognized earlier have been designed out of the product or that suitable guards have been provided. Any remaining hazards that are not correctable must be identified and suitable warning devices developed to alert the user to the existence, nature, and severity of the hazard. An indication of the probable consequences of not heeding the warning should be included.

It is hard to overemphasize the importance of proper hazard warnings on products. One of the most common allegations in product liability lawsuits is that the manufacturer failed to warn the user of the product of the existence of a hazard, thereby leading the user to think that the product was safe and free from hazards. After the user finds out the hard way, by suffering an injury, legal action is brought against the manufacturer, based on the "failure to warn" allegation. If the allegation is found to be true, in fact, the product may be judged to be defective and the manufacturer will be liable for damages due to this defect.

Warnings

The design and selection of warnings are far from a trivial exercise. In this author's opinion, this aspect of product design has been badly neglected in the past and is in great need of more critical attention. Although mitigating a hazard by issuing a warning is considered the least effective way of dealing with it (as opposed to designing it out or inserting guards), in many instances warnings are vitally important to the successful marketing of a product.

Three criteria must be met in the design of a warning. First, it must cause the user to be aware of the hazard; i.e., it must catch the user's attention. This can be done by placing the warning in an obvious place and making it big enough to be noticed. Colors and the size of the letters are important considerations, and graphics can be an effective way to portray possible consequences. Second, the warning must be understandable by the user, and this understanding must include the nature of the hazard. Third, the warning must convey the severity of the hazard to the user. This can be done by the use of certain characteristic colors: red is most associated with the most severe hazards, and yellow is used for less severe ones. In all cases, however, remember that the reason for the warning is the existence of an unprotected hazard and that the purpose of the warning is to modify the users' behavior so they will not be injured by the product.

There are published standards for the design of warning labels. The American National Standards Institute (ANSI) standard Z 535.4 sets requirements for wording and colors to be used, but other standards are issued by the Consumer Products Safety Commission (CPSC) and by the U.S. Department of Labor under the Occupational Safety and Health Administration (OSHA). In addition, many individual companies have in-house design standards which fit their particular needs.

Three special words are commonly used to attract the attention of the user:

- *Danger* is used when the hazard is such that severe injury or death will occur immediately upon contact with the hazard source.
- *Warning* is used when the hazard is such that moderate injury will occur immediately or that severe injury or death may eventually result from contact with the hazard source.
- *Caution* is used when the hazard is such that minor injury will occur immediately or that moderate injury may eventually result from contact with the hazard source.

Such other awareness-inducing words as *Stop, Attention, Look,* or *Notice* are not acceptable, but the use of a large exclamation point (!) is used as a hazard alert symbol and is placed ahead of any of the three so-called signal words. The standard also relates the colors of the warning to the severity of the hazard. The use of the word *Danger* calls for the letters to be white on a red background, *Warning* requires black lettering on an orange background, and *Caution* uses black lettering on a yellow background. All the combinations are intended to provide high contrast between the letters and the background for good legibility, even under adverse conditions. Significantly, no other color combinations are permitted for the signal words under the standard. However, any text wording can be either black on white or white on black. If a pictograph is used rather than words, the drawing must be black on a white background.

The size of the letters used in the text is also specified in the standard. The metric system, rather than printers' measure, is used; the signal words must be at least 3 millimeters high (9 points) and text words must be at least one-half that size (5 points minimum). There is no specification for the width of the lines, but boldface is generally considered a must for this application. Note that the ANSI standards are consensus standards and represent the minimal requirements acceptable to those who participated in the development of the document. It is not surprising, therefore, to find that some authorities disagree with the ANSI specifications and recommend that the letters be at least 12 points high to improve legibility.

Because warnings are intended to get attention, they must be designed in an aggressive manner. Passive shapes which feature rounded contours are not as effective attention getters as sharply angled shapes. Rectangles and triangles, being highly angular, are usually the best shapes for warning devices. Circles and ovals are the least desirable, and fancy shapes are not recommended.

The materials used for the warning devices must be selected just as carefully as for any other component. They should last as long as the surface to which they will be attached, and the method of attachment should be sufficiently secure to prevent their accidental loss in service. Life testing is usually required to make certain that these requirements are met. Testing should also be carried out to ensure that the warning is effective before the final design of the warning device is chosen.

OSHA regulations do not deal directly with warnings which are attached to a product as a part of the product. Their regulations, set forth in section 1910.145, relate to signs and tags which may be attached to the product by an employer to protect employees from a perceived hazard at a particular location. In some instances, such as industrial machinery and equipment, the signs may be included with the product by the manufacturer and should be considered as a regular part of the product by the designer. The wording of the OSHA requirement is quite general, in contrast to the very precise specifications of the ANSI standard. As an example, section 1910.145 (e)(2) states:

> (2) Nature of Wording. The wording of any sign should be easily read and concise. The sign should contain sufficient information to be easily understood. The wording should make a positive, rather than a negative, suggestion and should be accurate in fact.

Regulations issued by the Consumer Product Safety Commission (CPSC) are contained in title 16 of the *Code of Federal Regulations* sections 1000 to 1512. They cover such products as architectural glazing materials, matchbooks, walk-behind power lawn mowers, swimming pool slides, coal- and wood-burning appliances, and bicycles. The regulations are extremely precise and comprehensive. Any designer of such equipment must be aware of the content of the CPSC regulations and follow them assiduously. Placing a hazardous product on the consumer market will probably result in a recall action being required at great expense to the manufacturer.

Guards and Guarding

The use of guards on machinery has generated a substantial literature, largely because of the complexity of the subject and the great number of possible ways to solve the design problem. The basic legal requirement is set forth in OSHA regulations, title 29 of the *Code of Federal Regulations,* subpart O, machinery and machine guarding. Section 212 is the one most frequently cited and covers general requirements for all machines. The word *all* is the key item in this heading, for several of the following sections deal with the guarding of specific types of machines. For clarity and reference, the first two paragraphs are given here verbatim from the regulations.

> 1910.212 General requirements for all machines
> (a) Machine guarding -(1) Types of guarding. One or more methods of machine guarding shall be provided to protect the operator and other employees in the machine area from hazards such as those created by point of operation, ingoing nip points, rotating parts, flying chips and sparks. Examples of guarding methods are barrier guards, two-hand tripping devices, electronic safety devices, etc.
> (2) General requirements for machine guards. Guards shall be affixed to the machine where possible and secured elsewhere if for any reason attachment to the machine is not possible. The guard shall be such that it does not offer an accident hazard in itself.

Note the use of the word *shall* in the requirement for guards in the first sentence of paragraph (a). Use of the word *shall* makes the requirement mandatory, not optional. It says that the machinery *must* include guards. Any employer who exposes employees to unguarded machinery hazards risks a citation by OSHA. This citation will usually include an assessment of monetary penalties. Obviously, it behooves the designer of such machinery to include the guards as an integral part of the machine whenever possible.

In some instances inclusion of the guards is not possible, e.g., on mechanically powered metal-forming machinery such as stamping presses. Stamping presses, which do their work by applying large forces to materials via a set of forming dies, are sold by the manufacturer without any forming dies included. In that condition, the press cannot injure an operator who might get trapped in the press opening, since there are no dies present to come together and there is no nip point at what would normally be the point of operation. Because of the myriad of possible configurations of dies and ways of arranging them, it is generally recognized that it is not possible for the press builders to provide guards which would protect all possible configurations. Responsibility for provision of suitable guards therefore rests with the users of the presses. This is clearly stated in the standard which applies to such machinery, ANSI B11.1, *Safety Requirements for the Construction, Care, and Use of Mechanical Power Presses.*

The designer of machinery should be aware that any exposed moving part of the machine is a possible hazard. This should be part of the hazard recognition process in the early stages of the design process. The hazard should be dealt with by designing it out of the machine, if possible, or by providing suitable guards.

There are three areas in a machine where most of the dangerous moving-part hazards are present:

1. *The point of operation.* This is the location where the machine does its work on the workpiece.
2. *The power train.* This is comprised of moving parts which get the power to the point of operation and includes shafts, gears, pulleys, chains, cranks, cams, etc.
3. *Auxiliary parts.* This includes feeding mechanisms and other parts which move while the machine is in operation.

Point-of-operation guarding is treated in section 1910.212(a)(3)(ii) of the OSHA regulations as follows:

> (ii) The point of operation of machines whose operation exposes an employee to injury, shall be guarded. The guarding device shall be in conformity with any appropriate standards therefor, or, in the absence of applicable specific standards, shall be so designed and constructed as to prevent the operator from having any part of his body in the danger zone during the operating cycle.

Note that the guards are mandatory, not optional, just as in the first citation. Note also that the key words in the design of the guard are *shall prevent.* The guard must make it *impossible* for the operator to get into the area where the point-of-operation hazard exists.

Several design requirements for a suitable guard are implicit in the wording of the regulations. First, the guard must prevent any part of the operator's body from coming in contact with any moving parts which present a hazard. Although this was cited in the point-of-operation situation, it also applies to moving parts which are not located at the point of operation. This covers nip points in the power train where belts run onto pulleys, chains run onto sprockets, or gears mesh with each other. Ingoing nip points may also require guards to prevent portions of the operator's clothing (rather than the specifically mentioned *operator's body*) from getting caught and pulling the operator's body into the machinery.

As stated in the last sentence of subparagraph (2) of section 1910.212(a), "The guard shall be such that it does not offer an accident hazard in itself." This implies that the guard has no nip points or shear points and no exposed sharp edges or unfinished surfaces, which may cut the operator if accidental contact is made.

Subparagraph (2) also requires that the guards be attached securely, either to the machine itself or to some other place which will prevent casual removal of the guard. The attachment details are just as important as the details of the guarding device itself. In addition, it is often desirable to provide an interlock on the guards which prevents operation of the machine if the guard is removed for any reason.

Since most machinery must be serviced from time to time, for either lubrication or adjustments, the guards should be designed to interfere as little as possible with these normally expected activities. In many cases the lubrication system components can be located outside the guards so that the guards do not need to be removed during the greasing or oiling operations.

Although it is not mentioned in the regulations, obviously the guard should not interfere with the efficient operation of the machine to which it is attached. Any unacceptable interference is simply an incentive for the operator to remove, bypass, or nullify the guard, thus rendering it ineffective. An inoperative or defeated guard is worse than useless since its presence gives an unwarranted sense of security to the operator.

Experience has shown that moving machine parts should be protected from the accidental introduction of foreign objects. Obviously such a mishap can result in damage to the machinery, which is bad enough, but it may also cause the foreign object to be thrown back out of the machinery as a projectile which can strike and injure a person in the vicinity. Recall that the first sentence in section 1910.212 calls for protecting "the operator and *other employees in the machine area* from hazards."

Methods of Safeguarding Machinery

There are five major methods available for providing safety guarding for machinery and several options in each category. The major methods are mechanical guards, mechanical and electrical devices, guarding by distance or separation, material input and output systems, and external auxiliary equipment.

Mechanical guards, which are the commonest type, may be fixed, adjustable, or interlocked. The fixed mechanical guard is the commonest and most preferred type since it provides permanent prevention against accidental contact with hazardous machinery parts. Fixed guards should be used whenever possible. Adjustable guards are used where the mode of operation of the machine may be expected to change and it is necessary to accommodate for dimensional differences by adjusting the guard. However, once "adjusted," the guard should remain securely in the new position so as to act as a fixed guard. Several types of self-adjusting guards are often used on hand power tools. They are actuated by relative movement between the workpiece and the guard. The stock pushes the guard out of the way during the feed part of the operation, then permits the guard to close again after the feeding is completed. Interlocked guards are the next choice and are used where neither fixed nor adjustable means can be devised. Interlocking guards may be mechanical, electrical, pneumatic, or a combination of these. This type of guard prevents operation of the machine until the guard has moved into a position which excludes the operator from the danger zone or the point of operation. An essential feature of the design of an interlocked guard is that it must put the machine in a safe mode if the guard should happen to fail for any reason.

Mechanical and electrical devices, as distinguished from guards, may protect the worker in several ways. The device may stop the machine if it senses that the operator has entered the danger zone, pull the operator's hands out of the point of operation when the machine cycles, require that the operator keep both hands on controls which are remote from the hazard location, or insert a barrier gate between the operator and the danger zone.

Presence-sensing devices may use light beams and a photoelectric cell to stop the machine if the light beam is broken by any intrusion or a radio frequency electromagnetic field which is disturbed by the capacitance effect of the operator's body entering the field.

Pullbacks are devices which are usually attached to the operator's wrists or hands by cords, and the cords are pulled by the mechanical operation of the machine, thus removing the operator's hands from the danger zone. These devices are not popular with workers since the devices restrict workers' movements and have the unwelcome psychological effect of making the operator feel tied to the machine.

Duplex controls arranged in series are often used to ensure that the operator has both hands out of the hazard area during the operating cycle of the machine. It is possible, however, for the operator to tie down one of the controls, thereby defeating the safety feature. To preclude this possibility, two-handed controls incorporate a time-delay feature which requires that both buttons be actuated within a preselected time internal. An excessive time, such as would pass if one button were tied down, renders the machine inoperative.

Another version of control devices is the trip bar or other force-sensitive detection means. These are usually arranged as an electromechanical device which cuts off the power to the machine if the operator leans or presses against the device. If installed in front of an ingoing nip point, such as on rolls or calendars, these devices must be located so as not to be subject to accidental actuating by the material being processed. An additional consideration in the use of trip bars is that the machine may continue to run after the power is cut off unless a brake is included on the driving motor to stop the machine quickly.

Gates are simple mechanical devices which are opened and shut by the motion of the machine during its operating cycle. Many examples of such devices are found on injection molding machines where a horizontally sliding gate in front of the point of operation protects the operator from the high pressures of the molding process. Gates are often interlocked to prevent operation of the machine with the gate secured in the open position.

One of the simplest and most effective types of guarding is the use of distance to separate the operator from the hazard zone. This is a question of location that must be addressed by the designer. The distances and sizes of openings permitted at the given distances are given in table O-10 in section 1910.217(c)(2)(vi) of the OSHA regulations pertaining to mechanical power presses. The dimensions listed here are taken from ANSI standard B11.1, *Safety Requirements for the Construction, Care, and Use of Mechanical Power Presses.*

Distance of opening from point of operation, inches	Maximum width of opening, inches
0.50 to 1.50	0.25
1.50 to 2.50	0.375
2.50 to 3.50	0.50
3.50 to 5.50	0.625
5.50 to 6.50	0.75
6.50 to 7.50	0.875
7.50 to 12.50	1.25
12.50 to 15.50	1.50
15.50 to 17.50	1.875
17.50 to 31.50	2.125
Over 31.50	6.00 maximum

The dimensions of the openings are chosen to prevent the fingers on average-size operator hands from reaching the point of operation in a mechanical power press.

Feeding and ejection systems for the material being processed can also be designed to promote operator safety if the designer takes a systems approach to the problem. One of the more rapidly advancing methods for doing this involves the use of mechanical feeding and parts retrieval devices, commonly called *robots*. In addition to the ability of robots to perform very monotonous repetitive tasks ad infinitum, the use of robots keeps the operator completely out of the danger zone at all times. This is the ultimate solution to the operator safety problem. In addition, robots can work for extended periods in areas where there is a high noise level or where the temperature is higher than that tolerable by the average worker. One limitation on the use of robots is that they may, under certain circumstances, create a hazard by striking a bystander. This hazard can be avoided by the installation of suitable fences or barriers around the robot or by the use of presence sensors to stop the robot when there is an intrusion into its operating space.

Full automatic and semiautomatic systems for feeding material to machines have existed for many years. Material which comes in long strips is commonly fed from a reel of material by means of an indexing device. Sheet stock, such as paper for printing, is fed by pneumatic or mechanical devices. Three-dimensional material can be fed through chutes which position the stock for the operation to be performed in the machine. All these devices keep the operator out of the danger zone except when there is a misfeed or hangup in the system. Fixed barrier guards can be used to prevent the operator from getting into the danger zone during the operating cycle of the machine.

Automatic and semiautomatic systems are also available for the ejection of the finished workpiece at the completion of the operating cycle. Many of these depend on a blast of air to lift the finished piece from the machine and blow it into a receiving bin; others use a mechanical retrieval device which is actuated by the operator. Both methods keep the operator out of the hazard zone; however, the air ejection systems often make loud sounds which may exceed allowable noise levels. In addition, they can cause debris or material chips to fly into the air and strike bystanders.

External auxiliary equipment, the fifth of the available methods for protecting the operator, is usually used in connection with other protective methods to give an added measure of safety. In the largest class of such devices are the *awareness barriers*. These barriers are not physically sturdy enough to prevent a person from passing through them who was intent on doing so, and often they consist of an elastic cord or a length of brightly colored plastic tape with a suitable signal word printed on it. All they do is to alert the person to the hazard. They do not prevent people from exposing themselves to the hazard after passing through the barrier. Because of this limitation, awareness barriers are not acceptable as protection from a continuing hazard which should be avoided by other more substantial means.

Shields and screens, both fixed and portable, are commonly used as protection against splashing fluids, flying sparks generated by grinding or spot welding operations, and the bright light generated by arc welding.

In material forming operations using presses, external safety aids consisting of part-holding tools are commonly used when manual methods are employed to put the workpiece into the press and to retrieve it when the cycle is completed. These devices are extensions of the operator's hands and serve to keep them out of the danger zone. In the event that the press makes a stroke during the loading or unloading part of the cycle, the holding tool may be damaged, but the operator is not injured. The tools are an example of the safety principle which calls for the operator to remain out of the hazard zone at all times.

Another example of operator aids is the use of push sticks during sawing or other woodworking operations. When making thin cuts on a table saw, the operator must guide the workpiece very close to the danger zone of the moving saw blade. To provide adequate separation between the operator's fingers and the blade, a push block is commonly used. This provides reasonable control over the workpiece to achieve the desired cut, but keeps the operator's fingers away from the danger zone.

As emphasized previously, the design process should include hazard recognition as an integral part of the design review procedure. The hazard recognition effort must include not only the foreseeable dangers during regular operation of the product but also any hazard which may be present during the servicing and routine maintenance of the product. For example, if lubrication must be done while the machine is running, the oiler must be protected from the hazards of moving parts by the provision of suitable guarding. If any unusual hazards are developed during servicing, warning signs should be provided to alert the repair workers to the existence and nature of the hazard. If the precautions that have to be taken are of a complex nature, some detailed explanation of the hazard and how to avoid it should be included in the service manual, and a reference to the manual should appear on the warning sign.

Because the need to guard exposed machine parts has existed for such a long time and the hazards have been so well investigated, an extensive literature has been developed on the subject. In addition to several ANSI standards for specific types of machinery, the National Safety Council has issued several publications relating to guards and guarding. Among these are the *Accident Prevention Manual for Industrial Operations*; *Guards Illustrated*; and *Guide to Occupational Safety Literature*. OSHA has published several pamphlets on the subject, including 2281, *Beware of Machine Hazards*; 2247, *Machine Guarding*; Bulletin 2057, *Principles and Techniques of Machine Guarding*; and Bulletin 3067, *Concepts and Techniques of Machine Safeguarding*. With the principles of safeguarding machinery so well established and the extensive literature so readily available, there is no excuse today for the production of machines which fail to include proper hazard guards.

Design Review 2

After all the detail design has been done, the models have been built and tested, and the product has been proved to accomplish its intended function, a second design review should be carried out. Just as for the first such review, all major departments of the organization should be present to assess and evaluate the product from their particular viewpoints. Expect many changes in the proposed design as a result of this review. Once these changes have been incorporated into the design and the effects of the changes have been evaluated, the formal design portion of the project has reached its final stages. The product is ready for release to production.

Design considerations do not end with the release to production, however, since methods must still be developed to ensure that the product is produced in accordance with the approved design. This implies the design of suitable inspection equipment and the writing of procedures to be used by the quality assurance department, including the establishment of acceptable variations from the ideal product.

The preparation of operator's manuals and service instructions should be carried out concurrently with the final design process so that these materials can be ready when the first products come off the production line. Close liaison should be maintained between the technical writers who prepare the manuals and the designers who worked on the hazard recognition portion of the project. Suitable signal words should be included in the text to alert readers to the existence of possibly hazardous conditions which may develop during the use or servicing of the product.

Packaging

As soon as the product design is sufficiently far along, the design of the packaging for the product should be started. Packaging design is a highly specialized field in itself, which must take into account many factors, including preservation of the product, protection of the product against damage in shipping and storage, and external markings. Military and government products usually have packaging and marking requirements spelled out in the product specifications, but the design of packaging for consumer and commercial products is commonly the responsibility of the manufacturer.

There are many standards, regulations, and specifications for packaging issued by various agencies. The designer should be familiar with titles 46 and 49 of the *Code of Federal Regulations,* which deal with shipping and transportation, respectively. Consumer products are subject to many regulations regarding marking of possible hazards. It is specifically recommended that the designer of packaging for consumer products consider any hazards which may be involved in removing the product from the package. The consumer should be made aware of any contents of the package which have sharp edges or other dangerous attributes which could possibly cause injury while being removed from the package.

Life-Cycle Considerations

In recent years, environmental considerations have been receiving greater attention than ever before. Special attention has been paid to problems created by the disposal of products at the ends of their useful lives. These factors should be addressed during the initial design stages of any product, but some are particularly difficult to foresee. An example is the household refrigerator, which typically lasts for 20 years. It is a large, bulky, and heavy item which is difficult to dispose of. Many people simply left them around their property. How could the designer, 20 years before, foresee that young children would find a discarded refrigerator an attractive place to play in and would crawl inside, close the mechanically latched door, and find themselves trapped, unable to open the door from the inside? Yet that did happen. So many youngsters were found suffocated in discarded refrigerators that laws were passed requiring that the doors be removed after disposal. More recent designs use nonlatching closures with magnetic strips to effect a seal around the doors, thereby designing the hazard out of the product.

Summary

The product design process begins with the recognition of an unmet need or other marketing opportunity. The first step is the formulation of a concise statement of the design problem to be solved. Proposed solutions are then developed, and the most promising one is chosen for further consideration. Detail design is then carried out, followed by a design review. Hazard recognition is a fundamental part of the design review process, reflecting consideration for safety of the ultimate user of the product. Recognized hazards can then be designed out, prevented by use of guards, or warned about in the product. A second design review should be done at the completion of the final design and testing phases of the project. Preparation of instructional materials which sufficiently stress safety of use or operation of the product is a vital part of the design effort. Although this is sometimes difficult to do, designers should consider the hazards presented by the product after it has reached the end of its useful life. Designers should be aware of the extensive literature presently available regarding the requirements for a reasonably safe product.

5
Design of Consumer Products

According to the economic reporting services, consumer spending accounts for about 60 percent of the nation's economic activity. It is by far the largest single segment of the U.S. economy, which explains why indices of consumer behavior are so closely watched by government and financial agencies. Its sheer size, coming from consumers buying products, also accounts for the large number of personal injuries which take place.

During the latter half of the 1960s, the number of injuries from consumer products reached such a high level that various consumer organizations exercised their constitutional right to appeal to Congress for redress of what they considered to be a grievance. In 1967, Public Law 90-146 was enacted, authorizing the establishment of the National Commission on Product Safety (NCPS). Congress charged the NCPS with finding out the extent to which U.S. consumers were being exposed to product hazards which were so severe as to be considered unreasonable. Their report, issued in 1970, showed that, indeed, millions of people in the United States were being injured each year by consumer products. As a result, Congress enacted Public Law 92-573 which is known as the *Consumer Product Safety Act.* It became effective at Christmastime in 1972.

To regulate consumer products, the law set up the *Consumer Product Safety Commission* (CPSC) as an independent agency to accomplish four objectives:

- To protect the public against unreasonable risks of injury associated with consumer products
- To assist consumers in evaluating the comparative safety of consumer products
- To develop uniform safety standards for consumer products and to minimize conflicting state and local regulations
- To promote research and investigation into the causes and prevention of product-related deaths, illnesses, and injuries

The way in which the third objective is implemented is very important to the designers of consumer products. Under section 7 of the act, the CPSC has the power to prepare and publish consumer product safety standards. The standards establish the requirements for consumer products and are based on performance rather than specifications. The requirements cover performance, composition, contents, design, construction, finish, packaging, warnings, and instructions. Obviously, CPSC is a very powerful agency when it comes to the design of consumer products.

Definition of Consumer Products

From their own wording, *consumer products* are those which are used by their ultimate consumer, i.e., the person who uses or consumes the article in question. This is in distinct contrast to *industrial products,* which make use of machinery or other equipment in their use or operation, and to *commercial products,* which are intended to produce revenue for their owners or operators. Usually the consumer is the end user of the product, there being no further modification or processing involved other than some assembly work. Consumer products are usually purchased locally from a retailer by an individual. No purchasing agents or persons trained in buying things are involved. There are seldom any specifications involved in the purchase. Purchasers rely on brand names, and there is little commingling of products. The prevailing practice is *caveat emptor*—let the buyer beware. "What you see is what you get" is the motto of many consumer product purveyors.

Consumer products are usually purchased in small quantities, in contrast to the bulk purchases of commercial or industrial products. Consumer products are distributed through a long channel with many steps between the manufacturer and the consumer, who is at the end of the chain. There is often a heavy amount of advertising involved in getting the consumer to buy a particular product in the competitive marketplace.

Consumer products are usually used in or around the home, in a residential or social setting rather than in a workplace environment. Users of the products may be any age, sex, or physical condition and may have widely varying educational, cultural, or economic backgrounds. Because of this lack of homogeneity, the markets for consumer products are usually highly specialized and fragmented into small niches, with each producer fighting for her or his share of the market.

Because of the extremely wide spectrum included in the broad definition of consumer products, the Consumer Product Safety Act has a specific definition of its own. In the language of the act, a *consumer product* is any article, or component thereof, produced or distributed for sale to a consumer for personal use, consumption, or enjoyment in or around a permanent or temporary household or residence, a school, in recreation or otherwise. There is an extensive list of products which are *not* covered by its provisions, primarily because they are regulated by other federal laws. These include food, drugs, cosmetics, boats, aircraft, tobacco products, motor vehicles and their equipment, pesticides, and firearms.

Banned Products

One of the more draconian provisions of the Consumer Product Safety Act allows the CPSC to ban certain products from commercial distribution in the United States. This prohibition can be applied whenever the CPSC determines that the product presents an unreasonable risk of injury and that no feasible product safety standard could be developed which would provide adequate protection against the recognized unreasonable hazard. In addition, the CPSC can order the seizure of products which it considers to present an "imminent" hazard to consumers. Such hazards threaten immediate and unreasonable risk of death, serious illness, or severe personal injury.

Products which have been banned from commercial distribution include the following items:

1. *Unstable refuse bins.* In June 1978 metal bins of 1-cubic-yard capacity or larger which could tip over and crush young children were banned. About 700,000 bins had to be taken out of service or modified to meet stability requirements.

2. *Extremely flammable contact adhesives.* In January 1978 contact adhesives which had a flash point of −20°F or lower were banned be-

cause of the extreme risk of accidental ignition of the vapors from the solvent. Other contact cements of equivalent performance but without the flammability hazard were available as substitutes at the time of the ban.

3. *Lead-containing paint.* After February 1978 paint having 0.06 percent or more of lead content in the dried film and consumer products such as toys or furniture finished with such paints were banned. The unreasonable hazard was lead poisoning resulting from very young children eating bits of the lead-containing paint.

4. *Asbestos-containing patching compounds.* These were banned after June 11, 1978, because the loose asbestos particles could be inhaled, causing respiratory disorders.

5. *Lawn darts.* These were banned after December 1988. Lawn darts were 4- to 8-ounce metal shafts about 12 inches long and fitted with fins along their rear portions. They were used outdoors, being launched upward with an underhanded throw to land on a target area located several feet away. The unreasonable hazard was that the dart, which had a dull point, would fall onto the head or shoulders of a person, causing a severe puncture-type injury. Prior to the ban, more than 600 such injuries were reported each year.

The imminent-hazard provisions of the CPSA have been used rarely. In 1988 the hazards of lead in drinking water were recognized, and Congress enacted the Lead Contamination Control Act. The administrator of the Environmental Protection Agency (EPA) prepared a list of all drinking water coolers which had lead-lined tanks. These products were considered to be imminently hazardous consumer products, and the CPSC was directed to issue a recall notice to the manufacturers and importers of all such products.

CPSC Product Safety Standards

Since the Regulatory Reform Act, the number of bannings has dropped drastically. The preferred method of regulation has turned to the development of CPSC standards, using inputs from industry, safety organizations, technical societies, and government agencies. A classic example of this procedure is the *Safety Standard for Walk Behind Power Lawn Mowers.* This standard became effective for all rotary walk-behind power mowers on December 31, 1981. After that date, all such mowers had to be designed and manufactured in accordance with the standard. Otherwise, the product was declared to be unlawful. Significantly, the design requirements have been incorporated into standards published by the American National Standards Institute (ANSI) as Standard B71.3.

CPSC safety standards have been published for architectural glazing materials such as storm doors, bathtub and shower doors and enclosures, and sliding-type doors used on patios. The objective was to reduce the hazards of injury from glass being broken as a result of human contact. Data available to the commission indicated that nearly 200,000 lacerations, contusions, abrasions, or other injuries were associated with architectural glass products each year before the standard became applicable in 1977. As for the power lawn mower requirements, the glazing standard has been published as ANSI Standard Z97.1.

Even the ubiquitous book of matches has come under scrutiny by the CPSC because of unreasonable risks of burn injuries, eye injuries caused by burning fragments, and fires started by extinguished matches which had a lengthy afterglow period. In this instance, the CPSC published the general requirements as mandatory design and manufacturing specifications for book matches produced after May 1978.

In addition to the promulgation of mandatory product design standards, the CPSC can require that manufacturers prepare and provide certain information to consumers regarding the installation, operation, and maintenance of their products. An example is found in the regulations applicable to wood- or coal-burning appliances, meaning wood stoves in particular. These consumer products became very popular for home heating uses after the drastic increase in energy costs in the middle 1970s. Unfortunately, many were not properly installed. They were placed too close to combustible surfaces and had a tendency to accumulate creosote and other combustible materials in their flues and chimneys. As a consequence, there were many fires. In the opinion of the CPSC, this situation presented an unreasonable risk of injury from fire.

The CPSC instigated action to correct this situation by requiring manufacturers and importers of coal- and wood-burning appliances to supply technical information and performance data for the appliance to the purchasers. The requirements are very specific and detailed. For example, a written notice, printed in clear, legible, and understandable English, must be posted on the appliance in a visible location. The notice must be made of materials which will ensure legibility for the maximum expected useful life of the appliance. In addition, instructions for installation and operation must be provided. The directions are to include detailed, step-by-step installation procedures, along with the consequences to be expected if the correct procedures are not followed. Designers of such equipment must pay strict attention to the CPSC requirements and should have copies of the applicable documentation, which is published in section 16 of the *Code of Federal Regulations* as part 1406.

Other CPSC Responsibilities

Prior to the enactment of the Consumer Product Safety Act in 1972, there were several other laws in effect which also related to consumer product safety but in a less general way. These were:

- The *Refrigerator Safety Act* (RSA) was enacted after several small children were suffocated when playing in discarded refrigerators. This act requires that latchable doors be removed from old refrigerators and that new models incorporate doors which can be opened from the inside.
- The *Poison Prevention Packaging Act* (PPPA) requires the use of child-resistant closures on household items which could cause harm if ingested by children.
- The *Flammable Fabrics Act* (FFA) was enacted to reduce the possibilities of accidental ignition of clothing, particularly children's sleepwear.
- The *Federal Hazardous Substances Act* (FHSA) includes the *Child Protection Act* and the *Child Protection and Toy Safety Act* as amendments.

Designers should be aware of some of the definitions in the FHSA because they are often found in other contexts but not precisely defined. For hazards, the following are important:

- *Electrical hazard.* An article may be determined to present an electrical hazard if, in normal use or when subjected to reasonably foreseeable damage or abuse, its design or manufacture may cause personal injury or illness by electric shock
- *Mechanical hazard.* An article may be determined to present a mechanical hazard if, in normal use or when subjected to reasonably foreseeable damage or abuse, its design or manufacture presents an unreasonable risk of personal injury or illness: from fracture, fragmentation, or disassembly of the article; from propulsion of the article or any part or accessory thereof; from points or other protrusions, surfaces, edges, openings, or closures; from moving parts; from lack or insufficiency of controls to reduce or stop motion; as a result of

self-adhering characteristics of the article; because the article or any part or accessory thereof may be aspirated or ingested; because of any other aspect of the article's design or manufacture.

While it is reasonably certain that the last item was thrown in by some lawyer trying to be comprehensive and complete in the definition, the vagueness of its wording should strike terror into the hearts of designers of mechanical articles.

As experience grows, Congress passes more laws in response to real or perceived problems with consumer products. It assigns them to the CPSC for administration and the preparation of suitable regulations. Two examples are the safety rule issued by CPSC regarding automatic garage door openers and the Fire Safe Cigarette Act.

The garage door opener legislation, enacted in November 1990, mandates that such devices incorporate an entrapment protection feature meeting the requirements of a standard developed by ANSI and Underwriters Laboratories. The CPSC thereupon issued a rule to that effect under its existing rule-making authority.

The *Fire Safe Cigarette Act* of 1990 resulted from previously authorized studies which found that cigarette-caused fires led the national fire death toll, with nearly 1500 fatalities annually. In addition, more than 3800 injuries and nearly $400,000,000 in property damage resulted from fires caused by cigarettes in 1987. It was further determined that it is technically feasible to produce a cigarette which has a significantly lower tendency to ignite furniture and mattresses. The CPSC was directed by Congress to do further research and development work on the problem. This effort was to include the development of a standard test method for determining ignition propensities of various cigarettes and a computer model of the ignition physics, to generate a predictability algorithm, and to collect database information regarding the types of products being ignited and the smokers involved in the fires. Of course, smoking in bed and leaving burning cigarettes lying around on furniture are entirely the fault and responsibility of the smoker, not the manufacturer. But the hazard from misuse or abuse of the product is a foreseeable one, and it has been recognized for many years. The CPSC is taking the approach of attempting to design the hazard out of the product. As stated previously, this is a sensible approach which may yield satisfactory results.

CPSC Databases

In addition to developing safety standards, Congress assigned the CPSC the objective of "promoting research and investigation into the causes and prevention of product-related deaths, injuries, and illnesses." To achieve this objective, the *National Injury Information Clearinghouse* (NIIC) was established to collect information on injuries and their causes. The NIIC relies on three primary sources for its information input. First, injury data are obtained from hospital emergency facilities located throughout the United States. Second, data on fatalities which may involve a consumer product are obtained from copies of death certificates submitted by officials from all 50 states on a confidential basis. Third, the NIIC maintains a file of consumer complaints, which cite actual injuries or potentially hazardous situations, and collects published media accounts of product-related mishaps.

The primary data source is called the *National Electronic Injury Surveillance System* (NEISS). It was established in early 1972 and was fully operational several months before the Consumer Product Safety Act became effective in December of that year. Since then several million injury reports have been transmitted from hospitals to the CPSC central computer for analysis. Each report contains information about the victim (age, sex, body part involved, diagnosis), date of admission, location of hospital, and treatment given. Each incident is assigned to one or more of about 1000 separate product classifications for compilation. Because of the very large size of the database, statistically valid conclusions relating products to injury causation can be drawn. Many CPSC actions are based on this information and analysis when they reveal a recognizable and correctable pattern.

When the data show a pattern of suspected unreasonable hazard in a product category, further and more detailed accident investigations are begun. Information regarding the sequence of events leading up to the mishap, the manner in which the product was being used, and the behavior of persons involved may be obtained from interviews with witnesses and official reports. If a showing of an unreasonable hazard is found, recommendations for CPSC action can be formulated. This action may be in the form of preparing and issuing rules or regulations, monitoring the effectiveness of existing standards, or replying to consumer complaints.

Designers of consumer products should be aware that CPSC publishes most of the results of their accident data collection and analysis. These are in the form of Hazard Analyses, Special Studies, and Data Summaries. These publications identify hazards and report accident patterns by types of products. With such exhaustively researched reference material readily available, there is no excuse for designers to fail to inform themselves of the safety requirements applicable to their particular product line. As pointed out previously, both manufacturers and industry groups work with the CPSC in the development of voluntary safety standards. Various consumer advocacy groups utilize CPSC data to substantiate their positions on issues, and attorneys for injured plaintiffs draw on CPSC accident statistics to add weight to any claims of faulty product design. All of the NIIC design information is available from the CPSC at 5401 Westbard Avenue, Washington, DC 20207, or by calling 301-492-6424.

Compliance and Enforcement

Section 15 of the Consumer Product Safety Act augments the banning authority of CPSC, which is established in section 8. As an enforcement measure, section 15b empowers the CPSC to require all portions of the product distribution function (manufacturer, distributor, wholesaler, retailer, jobber, importer) to inform the CPSC immediately if any of them becomes aware of a serious defect in their product. This compulsory self-policing power of the CPSC has resulted in more than 2300 voluntary recalls which involved nearly 200 million nonconforming products.

In addition to the voluntary-recall provisions of section 15b, there is an involuntary-recall authority in section 15d. More than 140 million articles have been taken off the market by over 1250 section 15 recall orders issued between 1980 and the middle of 1991.

CPSC Fact Sheets

As part of their public information program, the CPSC issues a set of Fact Sheets which are directed to the consumer audience. They give some background information about the nature of hazards connected with certain specific products and data on the frequency and severity of injuries which are causally related to that product. Often they include case histories of accidents. Always they include suggestions for selecting a product and precautions to observe when using or operating it.

New Products

As originally written in 1972, the Consumer Product Safety Act was intended to correct problems with products already on the market and to prevent problems with products that were in the development stages but had not yet been introduced to the marketplace. Section 13 of the CPSA gave the CPSC the power to require the manufacturer of any new consumer product to notify the commission of his or her intention to introduce the new product and to provide a description of the product for CPSC review. The definition of what constitutes a "new" product was quite broad. There are two parts: a *new consumer product* is one which incorporates a design, material, or form of energy exchange which has not previously been used in consumer products, *and* not enough information exists about the product to determine the safety of the product when used by consumers.

In practice, the CPSC found that this power was almost never invoked. The sheer number of new consumer products being introduced in the marketplace each year, particularly in the home computer and electronics field, was simply overwhelming. Since any new design would have required a CPSC review, the administrative burden could not be supported. In the spirit of deregulation, section 13 was repealed as unworkable.

Examples of Faulty Consumer Product Design

For almost a decade the author worked as a forensic engineer. One principal activity was the investigation of causes of accidents, mishaps, product failures, and other unfortunate occurrences which resulted in personal injury or damage to property. Work of this kind shows only the worst aspects of the various products. When there is a pattern of repeated incidents, it can lead to a cynical attitude toward product designers. The following are some examples of repetitive failures encountered in practice.

Women's Shoes

One of the commonest types of mishaps involved women who fell, tripped, or otherwise lost their balance when their shoes broke under their wearer's weight. The failure was usually one in which the heel suddenly separated from the sole of the shoe, causing loss of balance.

Shoe heels are attached to the soles by special barbed nails which have been heat-treated to increase their tensile strength. A common nailing method uses a set of five nails. They are driven downward through the rear portion of the sole in a conical pattern. The nails seat themselves in the upper part of the heel itself, thus making the attachment of the heel to the sole.

In the usual failure, the heel broke off in a direction toward the rear of the wearer's foot. The two nails along the front edge of the heel either pulled out or snapped in a brittle tensile-fracture mode. The other three nails then bent or snapped, permitting the heel to fold back.

Clearly this attachment method has proved inadequate to the task. While style and aesthetics play a major part in the design of women's footwear, the mechanical and safety aspects have not been given adequate attention. Some method of reinforcing the joint between the brow of the heel and the sole of the shoe is needed.

Sports Equipment

With the current emphasis on physical fitness, there has been a proliferation of sports equipment in the consumer market. A classic case of bad sports equipment design came to the author's attention during an investigation of a personal injury lawsuit.

The victim was a man in his late twenties who was playing in a recreational softball league. He had climbed out of the third base dugout and was walking toward home plate. A teammate was in the warm-up circle, taking some practice swings, using a practice weight which he had slipped over the handle of his bat. As he made a full swing, the practice weight suddenly slid off the end of the bat, made a line drive down the third base line, and struck the victim in the face.

Examination of the weight showed that it was a slug of cast iron in the shape of a doughnut. The hole was tapered to fit the slope on the handle of a bat. This was intended to allow the weight to slide down only to the point where the diameter of the hole matched the diameter of the bat handle. The whole slug was encased in a soft, rubberlike material which made the weight smooth to touch and would not scratch the finish on the bat. After the accident the covering was found to be pulled out of the hole, making the slug look like a plumber's sink plunger. Obviously the covering did not adhere to the weight and had separated from it.

So what caused this horrible mishap? Would you believe a case of mistaken identity? The investigation revealed that there were two different sizes of bat weights. The smaller and lighter of the two was intended for use on softball bats. Their cylindrical portion is smaller than that on hardball bats, and softball bats are several ounces lighter. To accommodate these differences, the manufacturer made the weights in two different sizes. The coverings were made of different colors to differentiate them. There were also instructions with the weights, telling how to put them on and get them off a bat in the proper manner.

The manager of the team made a mistake and bought the hardball-sized weight for use on his softball team's bats. The person who designed the weights had unthinkingly made the hole in the hardball weight just large enough to slip over the cylindrical portion of a softball bat. A foreseeable misuse of the product was missed by the designer, for a small change in the diameter of the hole could have designed the hazard out of the product.

Early versions of exercise bicycles were also examples of designs which overlooked foreseeable hazards. In several investigations it was found that the sprocket and drivechain guard did not fully enclose the bottom portion of the chain. In particular, the point where the chain fed onto the rear sprocket was exposed on the lower side. This exposed in-running nip point was a hazard just waiting for young children to stick their fingers into.

Most early-model exercise bikes used chain guards identical to those used on riding bicycles. There was no real need to cover the rear sprocket on a moving bicycle, so the nip point there was left unguarded. For a stationary bike, however, it is foreseeable that a young child may be in the same area as her or his parent when the bike is being used. On later models, this obvious hazard was easily designed out of the product by extending the guard to cover the rear sprocket.

Ladders

Here is another common consumer product which was fraught with mishaps for many years. The Consumer Product Safety Commission considered ladders a high-priority item when they established product hazard rankings in the late 1970s. However, the ladder manufacturers, through their trade association and the A 14 committee of the American National Standards Institute (ANSI), have developed reasonable design, testing, and certification standards which have dramatically improved the safety of these products. CPSC monitored the efforts of the ANSI committee known as A 14 as they worked to develop an acceptable standard. Yet even today, one of the leading causes of household injuries is falls from portable ladders. The Consumer Federation of America (CFA) data indicate that about 90,000 ladder-related injuries take place each year.

Design efforts of the A 14 committee centered on strength and stability. Three different classes of portable ladders were established, depending on their load-carrying capabilities and their intended use. Light-duty ladders for use around the home were called class I, while heavy-duty industrial ladders were given a class III designation, with medium-duty units in class II. Minimum dimensions were established for certain components, and performance tests were prepared to ensure structural integrity.

It was recognized early on that one of the most important factors in portable ladder safety was the human factor. Clear, definite, and unambiguous instructions for the proper use, care, and storage of portable ladders were prepared. Comprehensive warning labels were developed, with meticulous attention paid to the wordings and detailed requirements specifying where the labels were to be placed and how durable the labels were to be. In this author's opinion, portable ladders intended for consumer use are an excellent example of how product designs can be improved. Note, however, that some of the incentive for the improvement came from government agencies and consumer organizations, who were attempting to eliminate what they considered an unreasonable risk of injury from the marketplace. Cooperation rather than confrontation seems to have been a successful approach in this instance.

Power Tools and Equipment

This category of consumer products includes electrically powered hand tools for use inside the home and gasoline-powered or electrically powered equipment used outdoors. Woodworking tools such as circular saws, miter saws, and table saws are among the items most commonly used by do-it-yourselfers and home repair persons. Mowers, snow throwers, leaf blowers, and chain saws are favorites of those interested in lawns and gardens.

The principal hazard in these very useful and productive products is accidental contact with the working element of the tool. Obviously, if a saw blade will cut wood, it will cut flesh and bone, too. Furthermore, the cut is not a clean slice such as a scalpel produces, but results from a ripping and tearing action which removes tissue from the wound. Prevention of accidental personal contact with the cutting element must be a major objective of designers of all types of power tools.

To this end, major efforts have been devoted to the designing of guards. For years, portable circular saws had a guard over the upper half of the blade, but the lower half, which did the actual cutting, was commonly left exposed. If the saw was set aside with the blade still moving, the saw could "walk" rapidly and strike the user. The exposed blade could inflict serious injury. A guard was finally developed which used a spring-loaded visor to cover the lower part of the blade. The visor was pushed back around the periphery of the blade when the saw was placed against the workpiece. As soon as the saw blade came out of the cut, the spring returned the guard to the closed position. This solution to the guarding problem has been widely adopted for portable circular saws.

Miter saws are another story. They are stationary tools which make straight, angled, or beveled cuts across pieces of wood used for trim carpentry. The circular blade is mounted vertically in a carriage which can be rotated either left or right to generate the angle of cut desired. The cut is made by swinging the blade in a downward arc. The workpiece is supported on a horizontal table which has a railing along its rear edge. The rail keeps the workpiece from moving rearward while the cut is being made. It has a slot near its center through which the saw blade can pass regardless of the angle being cut.

The top portion of a miter saw blade is readily guarded by a fixed semicircular hood. The lower half presents a difficult design problem. For cuts which are nearly straight across, a pair of sliding or hinged side pieces can be placed on each side of the blade and extended slightly beyond its edge. For more sharply angled cuts, however, such side pieces strike the rear rail and prevent the saw from completing its downward swing. One solution has been to omit the right side piece. The theory is that a right-handed worker will use the right hand to swing the blade down, thus keeping it away from the blade. The left hand holds the workpiece and is protected by a side piece on the left side of the blade. Of course, this design leaves the entire right side of the blade exposed. In the author's opinion this is a completely unacceptable design because it poses a totally unreasonable risk of injury to the user. Sadly, miter saws with precisely this design have been made and sold in the United States. Fortunately, designers have been able to correct this glaring deficiency in later models.

Table saws in several sizes have been available to consumers for many years and have proved to be very useful tools. The blade projects upward through a slot in the cast-iron table, and the workpiece is passed over the blade to make the cut. Long ripping cuts are easily made with this tool, and it is a favorite with people who like to work with sheets of plywood. Guarding the blade of a table saw is usually accomplished by providing a hinged hood over the blade. The hinge allows the guard to lift up and slide along on top of the workpiece when it passes over the rotating blade. Such guards are quite effective, *but only* when they are in place. One chronic problem with table saw guards has been the strong tendency of users to remove them. Guards do make it hard to see the cut and are inconvenient to use if a bevel cut is being made. This is a challenge for designers: Make an effective and convenient guard for table saws.

Chain saws may be either electrically-powered or gasoline-powered. They are very productive tools when compared to alternative methods of cutting or pruning trees. However, their very potency can lead to serious personal injuries if the user loses control of the saw. Control can easily be lost if the saw chain catches on some resistant object, such as another tree limb or piece of chain link fencing. When the chain catches on something, it comes to a halt. However, the engine continues to drive the chain. The result is that the saw moves while the chain stands still. This condition is known as *kick back*. It can take place extremely quickly to the total surprise of the unsuspecting operator. The movement of the saw is upward and back toward the operator. As soon as the kickback has freed the chain from the obstruction, the chain begins to move again at once. The critical factor is that the movement of the saw is frequently large enough to bring the moving chain into contact with the operator's upper body. When that happens, a severe injury occurs.

Attempts to reduce the hazard from kickback have resulted in special profiles to reduce the hang-up potential for the chain teeth, brakes to stop the chain instantly if a hang-up does occur, hand guards around the handle of the power head, and protective tips for installation on the end of the chain guide bar. The manufacturers of chain saws have also provided warnings of the hazards of kickback and instructions on how to avoid it. These efforts have been effective for new saws, but many older models are still in service which need to be upgraded with new styles of chain and protective tips on the guide bars.

Lawn mowers used to be of the rotating reel and bed-knife construction that were pushed from behind. They occasionally threw something back toward the operator, but seldom with enough force to inflict bodily harm. Lawn mower injuries were of low frequency and severity. Then the late 1940s saw the introduction of power-driven rotary mowers. They cut the blades of grass not by shearing them at low speed against a bed knife, as the reel types did, but by hitting them so fast that the stems broke off cleanly from the mechanical shock load effect. As an example, for the commonly used 21-inch-diameter blade being driven by a one-cylinder gasoline engine at the usual speed of 3600 revolutions per minute, the tip of the blade moved at a speed of nearly 20,000 feet per minute. This is approximately 225 miles per hour. If the blade happened to strike a solid foreign object which was concealed in the grass, it could expel that object from the mower at high speed. Thousands of injuries were reported from this cause, many of them severe enough to require hospitalization for the victim. This thrown-object hazard was readily recognizable by designers, but it remained uncorrected for many years. The obvious solutions were to reduce the tip speed of the blade and to design the discharge chute to deflect objects in a downward direction.

Another readily recognizable hazard of rotary power lawn mowers is accidental contact with the moving blade. This could happen in several ways, all predictable. The most obvious case occurs when wet grass gets clogged in the discharge chute, and the operator puts his or her hand into the chute to clear the obstruction. In another scenario the operator pulls the mower backward over one of her or his feet. Both these hazards can be designed out of the product—the first by shutting off or declutching the engine if the operator leaves the normal working position, the second by providing a guard flap over the rear side of the mower deck. Both solutions have been applied successfully.

Because of the appalling number of injuries incurred by mower users, the CPSC placed such products high on its priority list for attention. The first of almost one hundred product safety fact sheets was devoted to power lawn mowers. It presented data on the number and nature of injuries, case histories, and suggestions for care to be taken in selecting and operating such equipment. Eventually a contract was awarded to Consumers Union to develop a standard for walk-behind power mower safety. This was issued as a mandatory standard in 1981. It was published as part 1205 of the Consumer Product Safety Act in chapter 16 of the *Code of Federal Regulations.* The requirements have been incorporated in ANSI B 71.3, *Safety Standard for Walk Behind Power Lawn Mowers.*

Power-driven snow throwers are another consumer product in which the early models incorporated a readily recognizable hazard that could be designed out of the product. The hazard was due to clogging of the discharge chute, a hazard already well known to designers of power mowers. Heavy, wet snow tended to clog in the chute, tempting the operator to reach in with a hand or push in a stick to free the obstruction. Various warning labels were placed on the chute, telling the operator not to do such things, but they were only partially successful.

Another related hazard was the clogging of the inlets of the machine. Helical augers were commonly used to break up snowdrifts and feed the snow into the discharge path. If the augers became clogged, the operator sometimes put his or her hand (usually with a glove on it) into the auger to free it up. Typically, the auger would begin to turn as soon as it broke free, and the operator's gloved hand would be caught and drawn into the mechanism. This is totally predictable and completely stupid behavior on the part of the human factor in the design. The obvious remedy for this glaring defect is to install a "dead-man" control for the power source. This will stop the drivetrain any time the operator has left her or his position. Even if a hand or stick were foolishly inserted in a snow thrower, the augers would not turn. Another ANSI standard has been issued by the B 71 committee to incorporate the dead-man control and other safety features into the design of snow throwers.

Toys

Although intended to provide a pleasurable and possibly educational experience for children, some toys are capable of inflicting harm. Two examples from injury investigations by the author are illustrative of predictable hazards which should have been revealed by a routine testing program.

Toy 1 was a double-barreled pop gun, modeled after a famous frontier weapon of days gone by. It had a pump-and-lever action to compress a charge of air and contain it with a cork. When the trigger was pulled, the air discharged with a loud bang. The expelled cork was caught by a fitting glued to the end of the barrel so that the pump action could recycle it for the next shot. The proud user of this toy was a 4-year-old. He had played with it ever since his mother bought it for him that afternoon. About suppertime he was hiding behind an easy chair, using the arms to conceal himself from imaginary attackers. He cocked the gun, snuck up over the arm of the chair, and fired. Just then his little brother pulled himself up above the edge of the opposite chair arm and caught the corks and the retaining fitting right between the eyes. The glued joint between the fitting and the gun barrel had failed.

Examination of the components of the toy gun revealed very poor assembly technique for application of the glue to the joint. The joint was only partially secured and was doomed to early failure. Clearly this design was defective since it depended on a glued connection to secure a potentially lethal component.

Toy 2 was a quantum-leap improvement over the prosaic water pistol. Instead of holding a few ounces, this water cannon could hold the better part of a quart. You could really soak your adversary with this baby, yessiree.

The toy consisted of a length of very high-quality rubber tubing. It was closed off at its lower end by a metal clip, and was fitted with a filling valve at the top. It was loaded under pressure by attaching the valve to the end of a garden hose. When the gun was full, the valve was closed and unscrewed from the hose, ready for action. Just crack the valve a little, and the pressurized water jetted out up to 35 feet away. But you could also get hit by the clip that held the bottom end closed, if you tried to fill the water cannon beyond its capacity. Overfilling was a thoroughly predictable thing for young kids to attempt. The designers' failure to anticipate this action resulted in the product's being withdrawn from the market with considerable loss to the manufacturer.

Believe it or not, safety regulations for toys are contained in the *Federal Hazardous Substances Act.* Laws such as the *Child Protection and Toy Safety Act* of 1969 and the *Toy Safety Act* of 1984 were combined with other laws and put under the purview of the CPSC. The regulations are published as subchapter C of chapter 16 of the *Code of Federal Regulations,* cited as 16 CFR section 1500. Toy designers must be intimately familiar with the published regulations since they carry the force of law.

Several kinds of products which could be used as toys were placed on the high-priority list by the CPSC. Among these were baby pacifiers; toys with small parts, sharp points, or sharp edges; skateboards; and bicycles. Several mandatory regulations came out of this classification effort. The following design features for toys are mandatory:

For use by children of any age:

- No shock or thermal hazards in any electrical toys
- No toxic materials in toys
- Severely limited lead in toy paints

For use by children up to age 3:

- Must withstand use and abuse, must be unbreakable.
- No small parts or pieces which could become lodged in the throat.
- Infant rattles must be large enough to not become lodged in the throat and cannot separate into small pieces.

For use by children up to age 8:

- No heating elements in electrically operated toys
- No sharp points or sharp edges on toys

In addition to the CPSC regulations, the toy industry has adopted several voluntary safety standards. These include placing age and safety labels on toys, making squeeze toys and teething rings so large that they cannot get stuck in an infant's throat, and limiting the length of strings on crib and playpen toys to 12 inches to prevent possible strangulation. In addition, chests for storing toys, while not toys in themselves, are required to have lids which stay open at any level to which they are raised. This is to prevent the lid from dropping unexpectedly and striking the child or trapping him or her inside the box. Designers should note that all these regulations and standards are directed to the elimination of easily recognizable and correctable hazards.

Stairs and Stairways

While one can debate if stairways are a consumer product, they are an article produced "for the personal use...of a consumer in or around a permanent or temporary household or residence...." As such, they fall under the CPSC definition. From the author's experience, stairs are among the more hazardous domestic articles. But just as for previously mentioned products, most of the hazards are easily identifiable and can be reduced or eliminated by proper attention to design details.

Detailed design requirements for both interior and exterior stairs have been spelled out in local, state, and national building codes for decades. In addition the human factors aspect of stairway design has been researched by the military and is specified in section 5.7.7 of MIL-STD-1472D. It is plain that stairway design has been studied in depth for years. So why are so many injuries attributable to stairways?

One of the primary reasons is lack of uniformity in the height of the risers and the depth of the stair treads. Riser height variation is particularly common at the top or bottom of a flight of steps. The bottom riser on exterior stairs is particularly troublesome if the ground surface that it meets has a slope to one side. Obviously it is impossible to provide a uniform riser height across the width of the stair in such a situation. The usual problem with the top step is that the riser height may be "fudged" to make it match the level of the floor to which it leads. In either case, the lack of uniformity in the height of the steps forces the person descending the stairs to break her or his stepping pattern. If this required break is unexpected, loss of balance can easily result and a fall can occur. The problem is not so severe in the ascending direction since the riser height variation is more readily visible while going up than while coming down.

Building codes address this problem by specifying that allowable variations in adjacent risers or treads be no more than 3/16 inch, with no more than a 3/8-inch difference between the highest and lowest riser or the widest and narrowest tread. It is easy to detect any such variation by sighting down the flight of steps along a line tangent to the noses of the treads.

Another major and preventable cause of stairway mishaps is the lack of suitable handrails. Experience—and it is abundant experience—has shown that loss of footing on stairways is highly foreseeable. Some device should be provided to permit the person to regain his or her balance in the event of a misstep. A handrail, properly installed at the correct height and of an easily graspable shape, is the obvious solution.

Inadequate overhead clearance on stairways is also an obvious and easily remedied cause of injuries, particularly those to the head. The prevailing standard calls for 6 feet 8 inches of vertical clearance above the tread nosing. This is the same as for a conventional doorway. This clearance will be satisfactory for all but the tallest persons going up a stair. There is a natural inclination to lean slightly forward when ascending, thus lowering the height of the head an inch or so. However, when one is descending, a reverse effect comes into play. The person still leans slightly forward to watch where she or he is stepping. This places the head in a vertical line above the next lower step while the body is on the step 7 inches higher. The result is that any person who is more than 6 feet 1 inch tall will have to duck or get a bang on the head. In the author's opinion, this is one instance in which a standard is inadequate and needs improvement in the next edition.

Summary

The design of consumer products is subject to many government regulations, state and local codes, and product standards which emphasize safety. Most of these documents have the force of law, and following their provisions and requirements is mandatory for the designer, not optional. As a result, one of the first things a designer must do in the initial stages of the design effort is to ascertain what laws, codes, standards, and regulations are applicable to the design problem.

The importance of hazard analysis in the initial design stages cannot be overemphasized for consumer products. Under the Consumer Product Safety Act, the consequences of introducing a product which presents an unreasonable risk of injury are extremely severe. In some instances the survival of the manufacturer may be threatened. It goes without saying that without a commitment by management to product safety, it will not happen. Thus the first step in ensuring the safety of consumers is a commitment by management as a matter of written policy. Implementation of that policy commitment can be done only through a suitable organization. This organization must include provisions for authority commensurate with responsibility and a program for instilling good safety practices at all levels.

Design reviews, hazard recognition, and risk reduction are essential parts of the safety enhancement effort. Concurrent with them, a program of documentation and change control must be installed within the engineering department to maintain the integrity of the safety effort. In the event that a purchased component causes product safety problems, control of purchased parts should be sufficiently strict that traceability is ensured to the basic sources. This includes vendor quality rating programs and proper keeping of sourcing records.

Obviously, the safety features designed into the product in the beginning must be maintained through the production process. This is commonly achieved by routine in-process quality assurance checks. At the end of the production line, the product must be packaged, handled, and shipped in a manner which will not degrade its safety aspects, and suitable operating, installation, or user instructions must be included. Even service after the sale is subject to CPSC control in that records are required of fixes, field repairs, reworks, and modifications which could affect product safety. The CPSC rationale for these requirements is that it permits a rapid and accurate location and recalling of any products found to present unreasonable risks of injury to consumers.

6

Design of Industrial and Commercial Products

Industrial and commercial products include the broad categories of tools, machinery, equipment, and accessories used to make other products. In contrast to consumer items, commercial and industrial products are intended to produce revenue for their owners. They are considered by accountants to be capitalized investments which are depreciated over a certain time, have a relatively long service life, and may be sold as used equipment or for scrap value. Commercial and industrial products wear out or become obsolete, but they are not "consumed" during their service life.

Industrial and commercial products are often bought to a prepared specification by professionals in the purchasing field. This is in sharp contrast with consumer items which are purchased by end users who are shoppers. Shoppers must select from what is offered at their local retail stores. Price and performance competition are limited to the consumer's local area market. Where there are several possible suppliers for a certain commercial item, however, it is common for the purchasing agent to ask vendors from any locality to submit a bid in accordance with the specification requirements. The lowest bidder is awarded the supply contract based on competition among the vendors. Some of the contracts are for amounts in the millions of dollars, in contrast to the usual consumer out-of-pocket purchases.

Whereas consumer products have a long distribution chain which includes the original manufacturer, distributors, wholesalers, and retailers, the industrial and commercial markets seldom use wholesalers and never use retail sources. In many instances, purchases are made directly from manufacturers through their authorized agents or representatives. Warranties, performance guarantees, and service contracts are prevalent.

One special class of industrial products comprises those purchased as components for another product. These components are installed by the purchaser when the product is originally assembled. For that reason the components are known as *original equipment manufacturer* (OEM) items. If those items have a short service life compared to that of the basic product, when they wear out, they are replaced with another similar or identical unit, called an *after-market* item. All the features of this distribution system have a profound effect on the responsibility of the various parties in the event of a product failure, recall program, personal injury incident, or property damage.

A key characteristic of commercial and industrial products is that they are used in a factory setting rather than in a residential or academic environment. Their users are almost invariably adults who are employed by the equipment owner. In that context, one of the major components in the equipment is the human being who operates it. The influence of human factors on equipment design is discussed in Chapter 3. Another pertinent distinction between industrial and consumer products is that some industrial products are custom-designed to meet the purchaser's performance requirements or specifications. In such instances, responsibility for design errors rests with the specifier rather than the vendor. However, if a performance-type contract is used, the vendor has design discretion for meeting the requirements and carries the responsibility for the decisions.

Regulatory Considerations

Massachusetts was the first state in the United States to pass laws pertaining to safety as a consideration in the design of industrial machinery. Many textile mills were located in Massachusetts, and the number of accidents and injuries in the mills reached a level which aroused social concern shortly after the Civil War. In 1867 a factory inspection law was enacted, followed in 1877 by a law requiring guarding for dangerous machinery. Other states followed, and in 1902 Maryland enacted the first workers' compensation law. However, it was declared unconstitutional in 1904. Federal employees were provided with compensation protection in 1908 by an act of Congress. It was not until 1911 that New

Jersey, Wisconsin, and Washington enacted workers' compensation legislation which withstood a court challenge.

One result of the workers' compensation laws was the assessment of insurance premiums against employers in proportion to the losses incurred. In that way an incentive was provided for employers to make the workplace safer, to keep their insurance premiums low. In 1906 the Russell Sage Foundation began collecting statistics on the frequency and severity of industrial accidents. This provided a database for organizations interested in improving safety for workers. Seven years later, in 1913, another fact-gathering group was formed which later became the National Safety Council (NSC). The NSC formulated recommended safety practices based on the experience data collected over a number of years and published the first edition of *Accident Prevention Manual for Industrial Operations* in 1946.

Although there were industrial codes and compensation laws in most states by the end of World War II, there still was no national law directed at improving safety in the workplace. In 1960 the first step was taken when specific safety standards were put into effect as part of the federal Walsh-Healey Public Contracts Act of 1936. Many industrial firms had become familiar with its terms when they undertook defense contracts during World War II, even though the law did not include safety requirements during those years. After 1960 all federal procurement orders required that the suppliers adhere to the safety standards. With this as a precedent, it was confirmed that compliance with the safety regulations was feasible.

In 1970 the federal *Occupational Safety and Health Act* (OSHA) was passed. A copy of this act is included as Appendix B. One of its goals was "to assure so far as possible every working man and woman in the nation safe and healthful working conditions." Administration of the act was assigned to the Occupational Safety and Health Administration (OSHA) in the Department of Labor, which published regulations to implement its provisions. These regulations were given the force of law by the following wording: "... Each employer shall furnish to each of his employees a workplace which is free from recognized hazards that are causing or are likely to cause death or serious physical harm to his employees and shall comply with occupational safety and health standards promulgated under this Act."

The regulations are published in title 29 of the *Code of Federal Regulations,* which is the title devoted to labor. Section 1910 gives the regulations for general industry, which is the portion most commonly applied to commercial and industrial products. The construction industry is covered in section 1926, and agriculture falls under section 1928.

Special industries, such as shipbuilding and longshoring, are in sections 1915 and 1918, respectively. When a general industry requirement is cited, it begins with 29 CFR 1910 followed by the particular paragraphs referred to. As an example, the paragraph pertaining to machine guarding is 29 CFR 1910.212.

Before OSHA came into being, it was recognized that the task of developing and writing the great number of necessary regulations would require considerable time. To expedite the implementation of the act, the administrators were allowed to draw on the large number of existing safety standards and adopt them as they saw fit for a period of two years. Such standards were drawn from several areas and formed the basis for the original version of the OSHA regulations. Consensus standards such as those published by the American National Standards Institute (ANSI) and the National Fire Protection Association (NFPA) were drawn on heavily.

Proprietary standards, prepared by experts and professional groups such as the American Society of Mechanical Engineers (ASME), American Society of Agricultural Engineers (ASAE), American Society for Testing and Materials (ASTM), and Underwriters Laboratories (UL) were the next most important source. While these standards lacked the consensus features previously mentioned, they did provide a solid technical basis for the preparation of regulations.

Unfortunately, obsolescence has overtaken many of the hundreds of standards which were adopted by reference when OSHA came into being in 1971. Even worse, these standards are still being used as the basis for the most current revisions of the regulations. The 1990 revision contains references to standards published in 1969 or earlier. Clearly, these references are obsolete, and many are no longer available. Designers should check the availability and applicability of the latest published standards rather than rely on the OSHA regulations alone.

In addition, many new standards have been published to accommodate products which did not even exist when the original OSHA regulations became effective. An example is powered industrial trucks. In section 1910.178 the regulations cover only low- and high-lift fork trucks, based on a 1969 standard. There is no mention of rough-terrain forklift trucks, guided industrial vehicles, or operator-controlled industrial tow tractors, all of which came on the market after 1969 and have their own applicable standards. Moreover, it is apparent that the OSHA regulations must be used with caution. Even though they are often obsolete, they still carry the force of law and must be respected.

In addition to the national OSHA, about half of the states have their own safety and health plans. Many of these use the OSHA regulations almost word for word. However, Michigan and the Pacific rim states

(Alaska, Washington, Oregon, California, and Hawaii) have regulations which differ in significant detail from the national model. Designers should be aware of possible pitfalls for products intended for sale in those areas.

As expected, once the original versions of the OSHA regulations were on the books, there were many attempts to correct errors, upgrade requirements, expand areas of coverage, and include new situations as they became apparent. Revisions of the regulations are made annually after a notice of the proposed revisions is published in the *Federal Register,* hearings are held, and the final version is adopted. It is simply common sense that designers use only the current edition of OSHA regulations when formulating their approach to a commercial or industrial product. Copies of the regulations are available upon order from the Superintendent of Documents, U.S. Government Printing Office, Washington, DC 20402.

To save time, many provisions of the voluntary standards used as references by the authors of OSHA were simply transliterated into governmentese. However, in some instances the safety provisions of the consensus standard were considered to be inadequate. This is not surprising since the voluntary standards establish only the minimum requirements and are frequently criticized for their lack of stringency. The authors of OSHA took it upon themselves to raise the level of protection provided to workers in several instances.

A prime example of this upgrading of previously published safety requirements lies in the regulations pertaining to mobile cranes. One of the hazards of these machines is that a person working behind the control cab cannot be seen by the operator. If the operator swings the cab around, there is a possibility that the worker could be struck by the swinging cab or caught between the bottom of the cab and the crawler tracks. This hazard was long recognized by the National Safety Council, data were collected and published, and a method of protecting against it by installing barriers around the rear of the crane was recommended. However, the published standard for mobile and locomotive cranes did not include a requirement for barriers when OSHA became effective. To correct this obvious deficiency, the authors of OSHA added it under their rule-making authority, and it became the legal requirement.

Detailed Design Requirements

The general intent of the OSHA regulations is to establish performance requirements from which designers can work out their own detailed solutions. An example is the wording used to establish the requirements for machine guarding in section 1910.212. This section gives General Requirements for All Machines and is quite broad in its applicability. The first paragraph states as follows:

> (1) Types of guarding. One or more methods of machine guarding shall be provided to protect the operator and other employees in the machine area from hazards such as those created by point of operation, ingoing nip points, rotating parts, flying chips and sparks. Examples of guarding methods are barrier guards, two hand tripping devices, electronic safety devices, etc.

Two things are noteworthy about the wording of this paragraph. First, it is quite general. It does not specify what type of guarding is to be provided, but leaves that decision up to the designer. Of course, there are certain generic classes of guards available, and some of those are mentioned, but because the details of each particular situation are unique to that application, no specific design details are given. Designers are left to use their own skills in their choice of guards.

The second notable thing about the wording of section 1910.212 is the use of the word *shall* in the requirement. The word *shall* makes the requirement mandatory. The designer *must* provide guarding. There is no choice. The regulation does not say *may* or *can* or *should.* Those words all imply that the designer has some discretion in the matter. The regulation says *shall,* and that word obligates designers to comply. Otherwise their product cannot legally be put into the stream of commerce. Note that the use of the mandatory wording style is common throughout the OSHA regulations, and designers who refer to certain applicable sections should be alert to its presence.

Although many OSHA regulations are general in their requirements, in many instances designers are given little or no latitude in the details. Often this is the result of heavy reliance on information published in established standards which have been adopted by reference or revised for incorporation into the regulations. An example of this situation arises in subpart D, which relates to walking and working surfaces. This subpart includes ladders and scaffolds of many types, along with stairways and railings.

For instance, fixed ladders are covered by section 1910.27, and that section is largely based on ANSI A-14.3, *Safety Requirements for Fixed Ladders.* The 1956 version of this standard is used as the OSHA reference, but this document has been revised several times since 1956, with the most recent revision in 1984. Designers should use the latest version. The design requirements are given in section 1910.27(a)(1). They include a live load of 200 pounds applied as a concentrated force at the location where it will develop the maximum stress in the member, but the allowable maximum stress is not specified.

Spacing between rungs is limited to 12 inches maximum, and the spacing must be uniform throughout the length of the ladder. The minimum clear length of rungs is given as 16 inches, but no maximum is given. The size of rungs is specified as ¾-inch diameter, implying that only round rungs are acceptable. Square bars set with one edge vertical do not comply with the regulation, even though they provide excellent traction when they are wet or icy. Side and rear clearances are also specified. The most important clearance, and the one most frequently overlooked by designers, is the minimum 7-inch toe clearance required along the rear side of the ladder between the supporting wall and the centerline of the rungs. This distance is required to ensure safety by permitting the climber to set her or his shoe deeply enough onto the rung to be able to rest the ball of the foot on the rung. This clearance requirement can be particularly important where there is an obstruction behind part of the ladder. In that case the rungs must be spaced so as to straddle the obstruction.

A particularly troublesome situation arises when the fixed ladder leads to a scuttle hole or hatch opening in a roof. The full 7-inch clearance must be provided between the top rung of the ladder and the inside face of the hatch opening. Since the ladder must extend clear to the top of the opening, often the thickness of the walls around the opening reduces the toe clearance behind the topmost rung. This makes it difficult for climbers to place their feet in a safe position both when climbing out of the opening and particularly when reentering the opening to go back down. Designers can eliminate this obvious hazard by paying proper attention to the details and following the regulations. Illustrations of the design requirements are given in Figs. 6.1 and 6.2.

As a contrasting example of nonspecific requirements, portable metal ladders are treated in section 1910.26. The regulations are largely based on the provisions of *Safety Requirements for Portable Metal Ladders*, ANSI A 14.2. As for the case of fixed ladders, the regulations are out of date with respect to current standards. They use the 1956 version of the standard as a basis instead of the most recent version, which was issued in 1982 with a supplement in 1985.

Because of the large number of suitable metals and the multitude of possible design approaches, no specific design requirements are set forth. The wording of the general requirements paragraph is so vague as to be almost meaningless. It states:

> The design shall be such as to produce a ladder without structural defects or accident hazards such as sharp edges, burrs, etc. The metal selected shall be of sufficient strength to meet the test requirements, and shall be protected against corrosion unless inherently corrosion resistant.

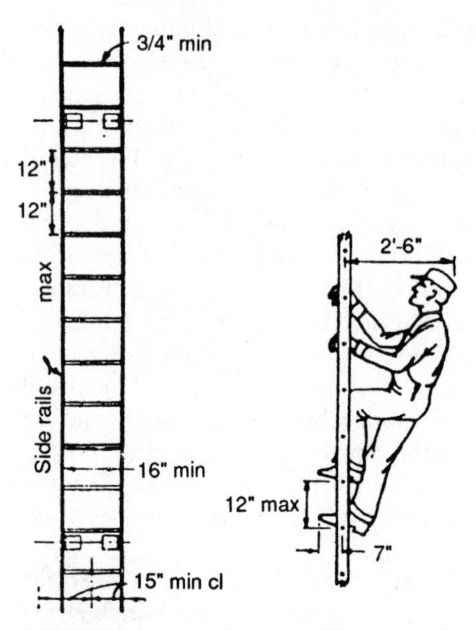

Figure 6.1. Rail ladder with bar-steel rails and round steel rungs.

Minimum ladder clearances

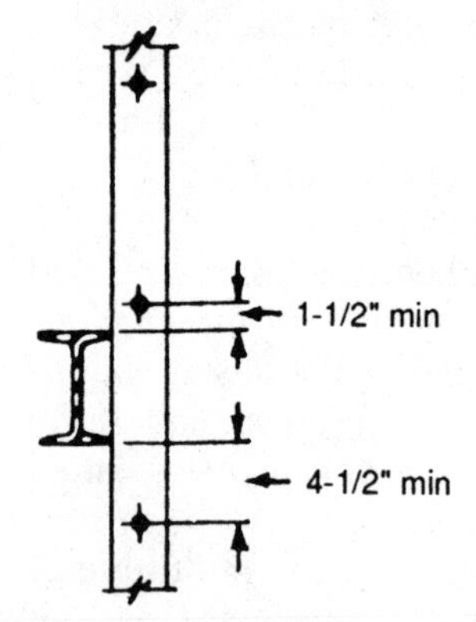

Figure 6.2. Clearance for unavoidable obstruction at rear of fixed ladder.

Clearly, ladder designers who are looking for an information source will be best served by obtaining a copy of the latest ANSI standard as a primary basis, then checking against the OSHA regulations for possible conflicts with the legal requirements.

OSHA Environmental Regulations

Since it can affect the health of workers, their working environment is subject to OSHA regulations in the same manner as the tools, machinery, and equipment which they use. The major environmental exposure of concern to designers of industrial and commercial products is the amount of noise their product generates. OSHA has promulgated very extensive and detailed regulations covering the amount of noise exposure allowed in the workplace. These are published in section 1910.95, "Occupational Noise Exposure."

The noise regulations are of the performance type and are not directed at any specific type of machinery or equipment. Instead they refer to the workplace itself without regard to the source of the noise. Inherent in the wording, however, is the implication that products which generate high levels of workplace noise are less desirable than those which are quieter.

This comes about through the wording of section 1910.95(b)(1), which states:

> When employees are subjected to sound exceeding those listed in Table G-16, feasible administrative or engineering controls shall be utilized. If such controls fail to reduce sound levels within the levels of Table G-16, personal protective equipment shall be provided and used to reduce sound levels within the levels of the table.

The words engineering controls are directed squarely at the equipment designers. The intent of the wording is to encourage noise elimination as early as possible in the product design stage. This approach is identical to the "design it out" philosophy used for the elimination of other personal-injury hazards. If designs cannot achieve the required sound level reduction, the next approach is to provide shields, mufflers, or sound absorbers as attachments to the equipment. This is analogous to the provision of guards for machinery. Obviously these added components increase the cost of the product and could place it at a competitive disadvantage in terms of price. The term *administrative controls* is a requirement that managers of the company using the noisy equipment schedule its use so that the allowable noise limits are not exceeded. This

Table G-16. Permissible Noise Exposures

Duration per day, h	Sound level, dBA slow response
8	90
6	92
4	95
3	97
2	100
1½	102
1	105
½	110
¼ or less	115

NOTE: When the daily noise exposure is composed of two or more periods of noise exposure of different levels, their combined effect should be considered rather than the individual effect of each. If the sum of $C_1/T_1 + C_2/T_2 + \ldots + C_n/T_n$ exceeds unity, then the mixed exposure should be considered to exceed the limit value. Here C_n indicates the total time of exposure at a specified noise level, and T_n indicates the total time of exposure permitted at that level.

Exposure to impulsive or impact noise should not exceed 140-dB peak sound pressure level.

implies shutting down the equipment after the time limit has been reached (thus causing a loss of production) or shifting the operators around so that no worker receives an overexposure to the noise. Obviously any equipment which is so noisy that it must be operated under administrative controls is not competitive with equipment which is not so encumbered. It is clear that the OSHA regulations provide several big incentives for designers to recognize the hazardous nature of noise generated by their products and to take effective measures to design that hazard out of the product.

Machinery and Equipment Design

There are three subparts of the OSHA regulations which deal directly with the design of machinery and equipment. Subpart N covers materials handling and storage equipment, including powered industrial trucks, several types of cranes, and derricks. Subpart O is devoted to machinery and machine guarding, including woodworking equipment, cooperage machinery, abrasive wheel machinery, mills and calendars used in the rubber and plastics industries, mechanical power presses, forging machines, and all sorts of power transmission apparatus. Subpart P gives requirements for hand and portable powered tools and equipment.

Subpart N

Regulations for powered industrial trucks are contained in section 1910.178. Manufacturers of such equipment should be aware that the design and construction requirements in the most recent OSHA regulations are hopelessly obsolete. They are based entirely on the 1969 edition of ANSI B56.1, *American National Standard for Powered Industrial Trucks.* While that was the applicable standard when OSHA went into effect on August 27, 1971, it has been superseded by ANSI/ASME B56.1-1988, *Low Lift and High Lift Trucks.* Clearly, designers should use the latest edition of the standard.

In addition, several new types of industrial trucks have been placed on the market since 1971 but are not even mentioned in the OSHA regulations. These include guided industrial vehicles under standard ANSI/ASME B56.5-1988, rough-terrain forklift trucks under ANSI/ASME B56.6-1987, industrial crane trucks under ANSI/ASME B56.7-1987, personnel and burden carriers under ANSI/ASME B56.8-1988, and operator-controlled industrial tow tractors under ANSI/ASME B56.9-1987.

Overhead and gantry cranes; crawler, locomotive, and truck cranes; and derricks are covered by sections 1910.179, 1910.180, and 1910.181, respectively. The basic OSHA design standard for overhead and gantry cranes is the 1967 edition of ANSI B30.2. This has been superseded by the 1990 edition of ANSI/ASME B30.2. The new standard applies to top-running trolley hoists and bridge cranes of either single- or multiple-girder designs. Crawler, locomotive, and truck cranes are now subject to ANSI/ASME Standard B30.5 which was issued in 1990. This replaces the obsolete OSHA standard of 1968. Similarly, the 1968 standard for derricks used by OSHA has been superseded by the 1990 edition of ANSI/ASME B30.6.

New standards have been issued for cranes not included in the OSHA regulations. These are ANSI/ASME B30.3, issued in 1990 for hammerhead tower cranes; ANSI/ASME B30.4, issued in 1990 for portal, tower, and pillar cranes; ANSI/ASME B30.8 for floating cranes, issued in 1988; ANSI/ASME B30.18, issued in 1987 for stacker cranes; and ANSI/ASME B30.22 for articulating boom cranes, issued in 1987.

Subpart O

Subpart O, which covers machinery and machine guarding, is one of the most important parts of the OSHA regulations for designers of industrial and commercial products. A set of general requirements which apply to *all* machines is given in section 1910.212. These requirements are based not on published consensus standards, but on safety provisions in the Walsh-Healey Act pertaining to public contracts. These items were in the *Code of Federal Regulations* long before OSHA came on the scene.

The primary emphasis in this section is on guarding of recognized hazards. Note that guarding a hazard is the preferred method only if the hazard cannot be designed out of the product. Such operator hazards as being pulled into in-running nip points, being caught on rotating parts, or getting trapped in the point of operation of a machine are typical opportunities for the use of guards. Shears, power presses, milling machines, power saws, portable power tools, forming rolls, and calendars are types of machinery where point-of-operation guarding is always called for.

The basic design principle for point-of-operation guards is that they must be constructed in such a way that it is impossible for any part of the operator's body to get into the hazardous zone at the point of operation at any time during the operating cycle of the machine.

Examples of in-running nip points commonly found in machinery are shown in Fig. 6.3. The nips between on-running belts and their mating pulleys are readily recognized, as are the nips between on-running chains and sprockets. The nips between the helical flites, helical metal parts that surround the shaft of a screw conveyor, and the trough on an auger screw conveyor are not so obvious, nor is the nip between the tool rest and the rotating grinding wheel. A really elusive nip is the one between the spokes of a handwheel and the guideway behind it. The clue for designers to recognize the existence of a possible nip point is the close proximity of one part to another, accompanied by relative motion between the parts which tends to reduce the separation between them. During the design review stages of the project, designers are advised to check the kinematic aspects of the design to discover any nip-point clues. This check is a basic part of the hazard recognition process.

Point-of-operation guards are located at the precise point where the machine does its work on the workpiece. The guard is there to prevent any part of the operator's body from unintentionally becoming the workpiece. Several examples of point-of-operation guards for power saws are shown in Fig. 6.4.

Figure 6.5 shows a power band saw as an example of a situation where it is extremely difficult, if not practically impossible, to provide a completely effective point-of-operation guard. While most of the exposed blade can be protected by an adjustable shield, the actual working opening must be exposed in order to permit the workpiece to pass over the moving blade of the saw.

In some instances the provision of suitable guarding on the machine itself is impractical, but the use of an additional item known as an *aux-*

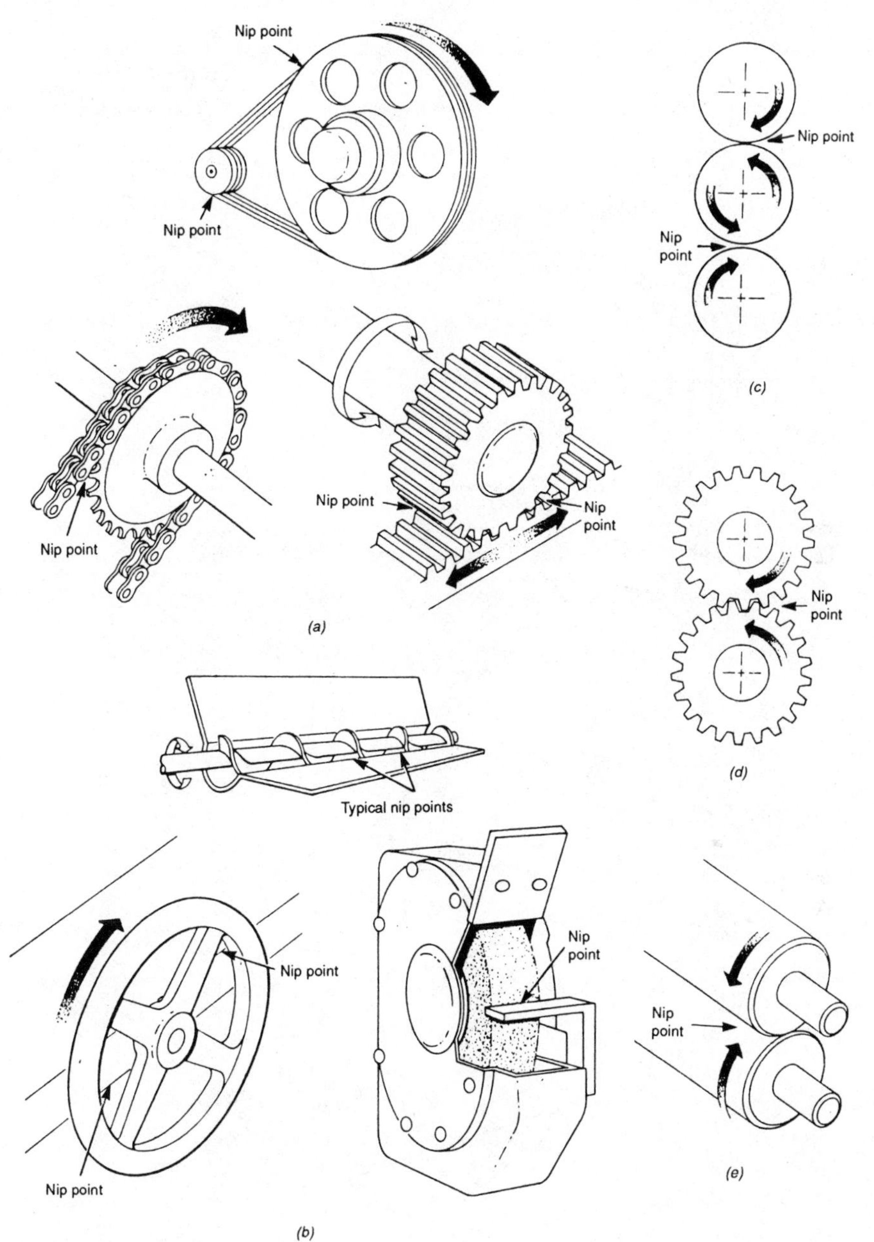

Figure 6.3. Examples of machinery nip points.

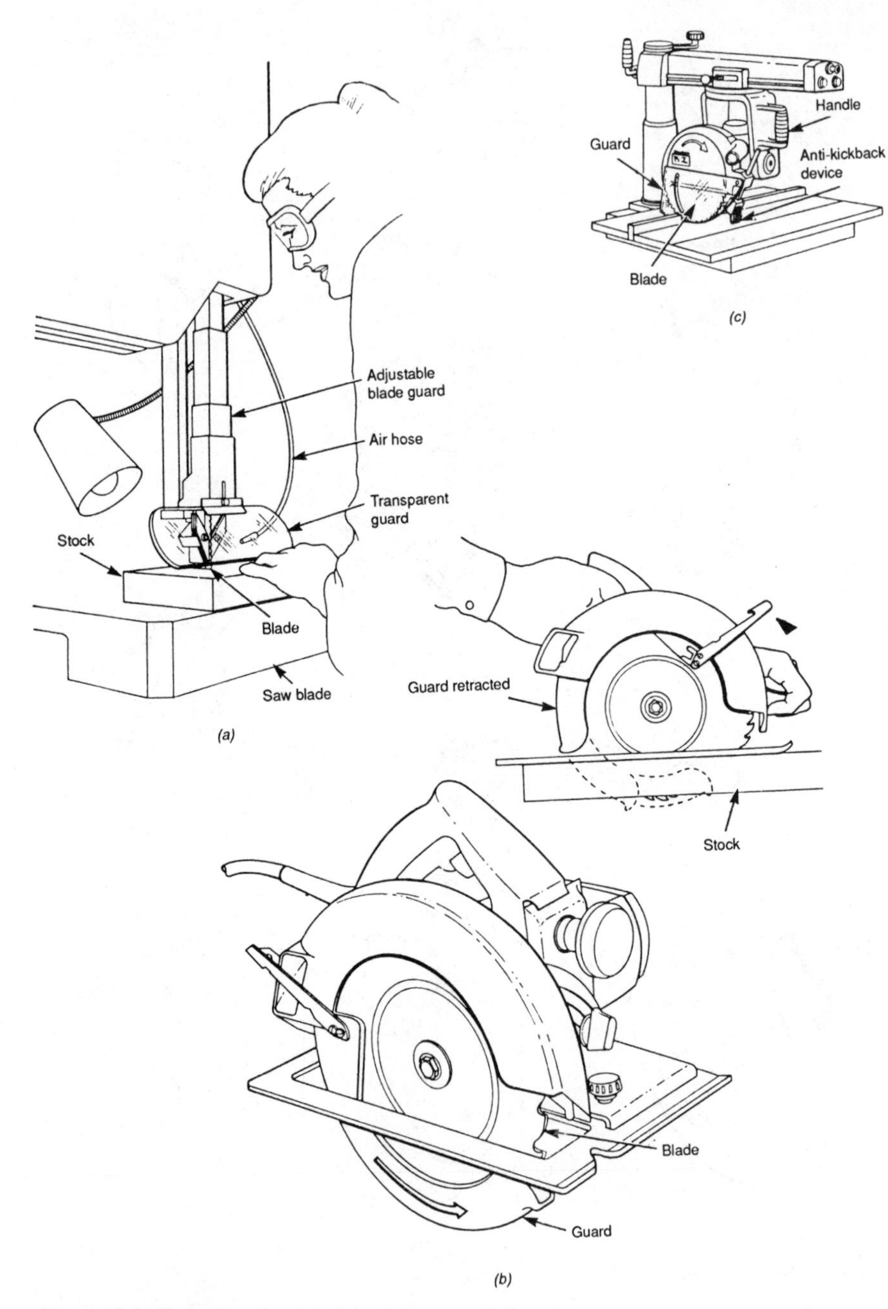

Figure 6.4. Examples of point-of-operation guards for saws.

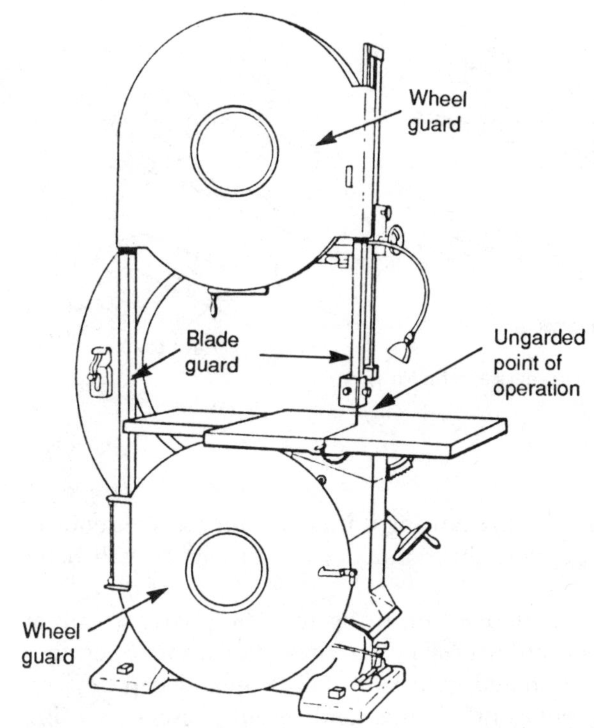

Figure 6.5. Examples of unguarded point of operation.

iliary device may solve the problem. For example, on table saws it is known that some types of cuts require operators to place hands in close proximity to the moving blade in order to feed the workpiece. In such cases it is common practice to provide an auxiliary device such as a push stick to avoid accidental operator contact with the blade. Two types of pushers are shown in Fig. 6.6. Variations on these designs can also be used on band saws where the point of operation is difficult to protect. Auxiliary devices are commonly used to load and unload workpieces from mechanical power presses. Designs for several such devices are shown in Fig. 6.7.

Mechanical power presses are the classic example of extremely durable industrial equipment used to produce other commercial and industrial items. Because of their very long service life, many power presses are in use which were built long before present-day safety requirements came into being. Many have been overhauled, rebuilt, and modified several times but are still productive capital assets. However, they carry some very serious hazards with them. OSHA found a way to reduce the extent of these hazards by requiring that anyone who reconstructs or

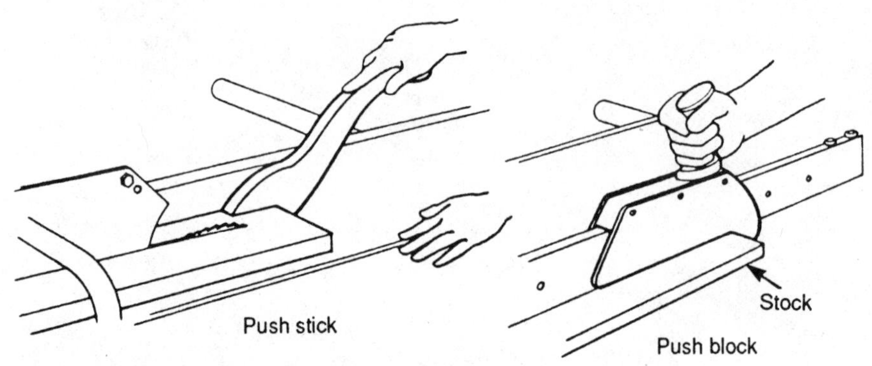

Figure 6.6. Two types of auxiliary devices for table saws.

modifies a mechanical power press bring it up to current safety requirements. Although well intended, this regulation has proved very difficult to enforce.

There are four basic methods of ensuring that the operator cannot get into the point of operation of a power press during its operating cycle: provide specially designed guards, install two-hand trip devices, attach pullbacks to the operator's hands, and install presence-sensing devices which prevent the press from making a stroke if the operating area is not clear. Illustrations of these point-of-operation guards are shown in Figs. 6.8 through 6.11.

In common with the reference standards used for other parts of the regulations, OSHA bases the subpart O requirements on obsolete documents. Woodworking machinery, which is covered in section 1910.213, is based on the 1961 revision of the 1954 edition of ANSI O1.1, *Safety Code for Woodworking Machinery*. This standard is no longer listed in the ANSI catalog. In 1983 it was partially replaced by ANSI O2.1, *Safety Requirements for Sawmills*. The 1970 version of ANSI B7.1, *Safety Code for Abrasive Wheels*, has been superseded by a 1988 version and supplemented with B7.5, *Safety Code for the Construction, Use, and Care of Gasoline-Powered, Hand-Held, Portable Abrasive Cutting-Off Machines*. The OSHA standards for mechanical power presses are based on the 1971 edition of ANSI B11.1, *Safety Standard for Construction, Care and Use of Mechanical Power Presses*. Since that time the B11 committee of ANSI published a whole series of standards

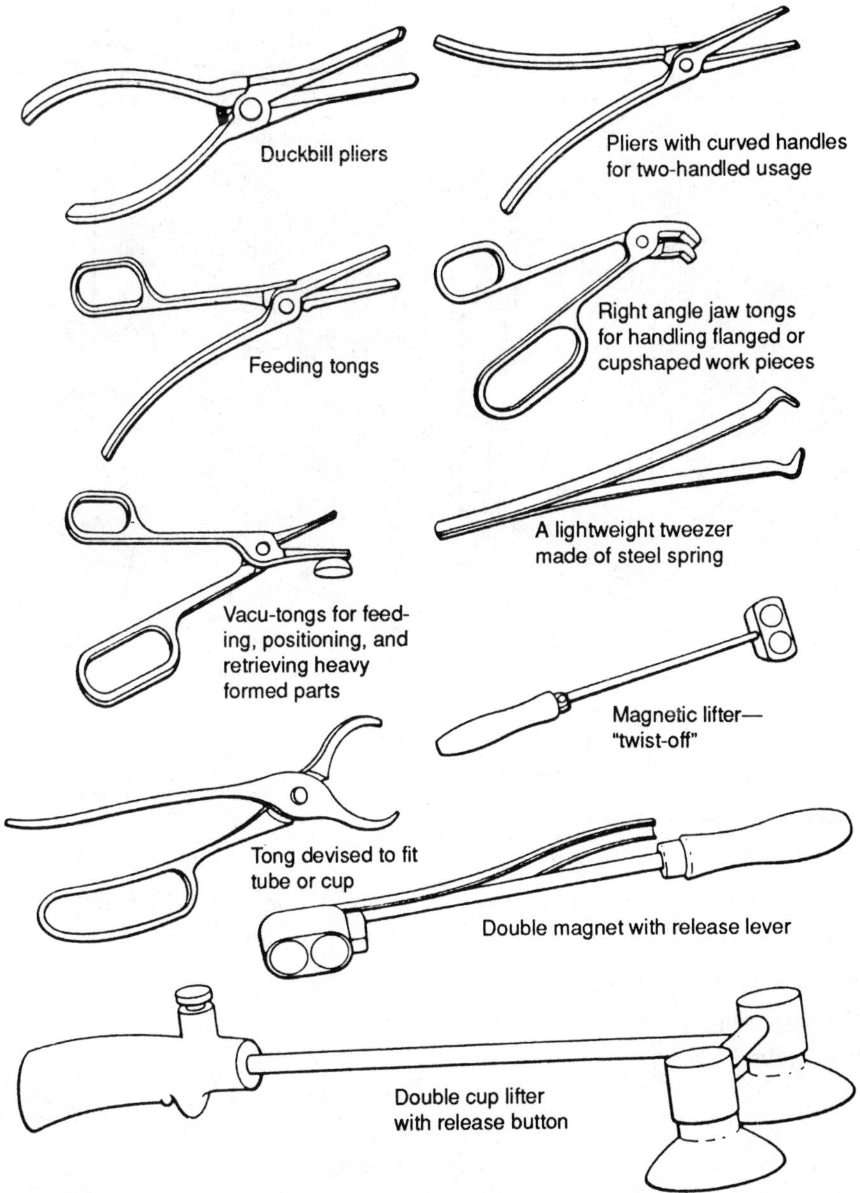

Figure 6.7. Types of auxiliary devices to load and unload workpieces from mechanical power presses.

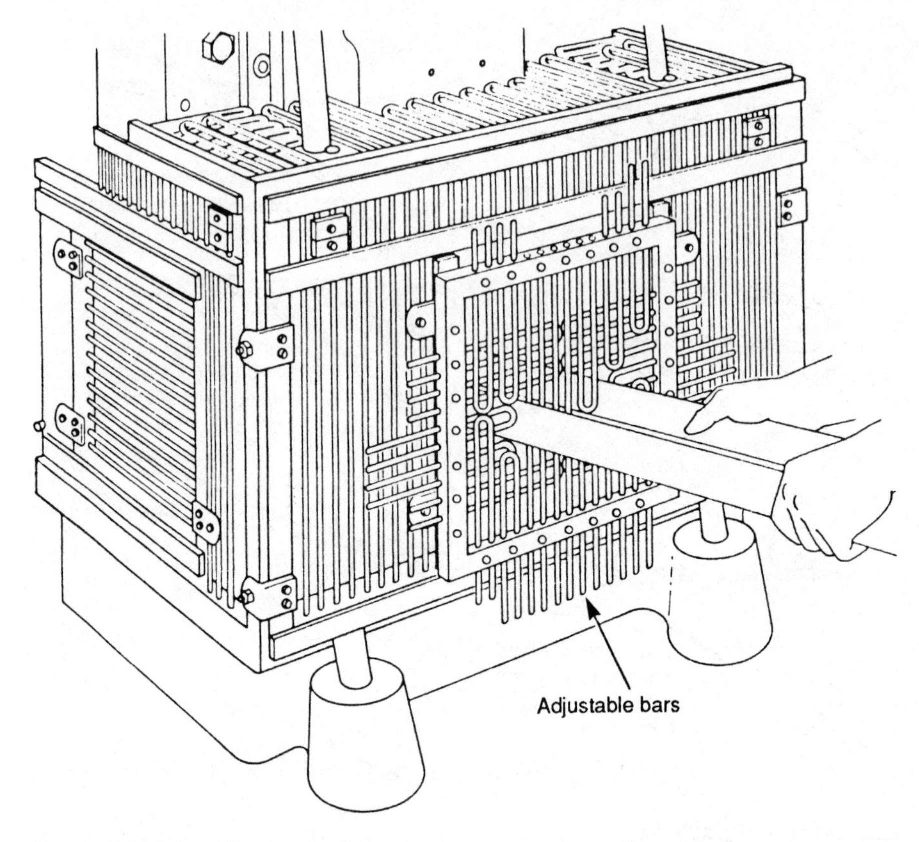

Figure 6.8. Adjustable cage guard.

relating to the entire field of machine tools. These cover the safety requirements for the construction, care, and use of the following:

Mechanical power presses	B11.1-1988
Hydraulic power presses	B11.2-1982
Power press brakes	B11.3-1982(R1988)
Shears	B11.4-1983
Ironworkers	B11.5-1988
Lathes	B11.6-1984
Cold headers and cold formers	B11.7-1985
Drilling, milling, and boring machines	B11.8-1983
Grinding machines	B11.9-1975(R1987)

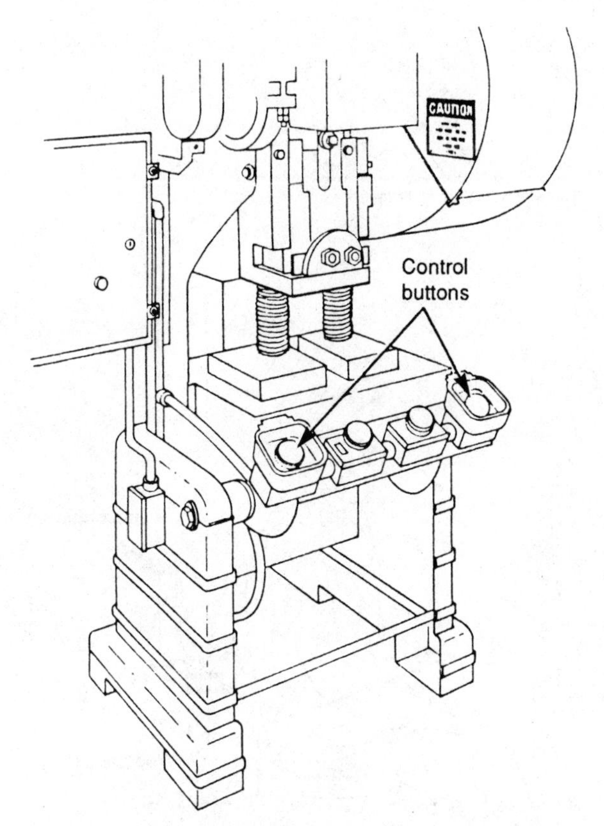

Figure 6.9. Two-hand trip press control.

Metal sawing machines	B11.10-1983
Gear cutting machines	B11.11-1985
Roll forming and roll bending machines	B11.12-1983(R1989)
Single- and multiple-spindle, automatic screw bar and chucking machines	B11.13-1983
Coil slitting machines	B11.14-1983
Pipe, tube, and shape bending machines	B11.15-1984(R1989)
Metal powder compacting presses	B11.16-1988
Horizontal extrusion presses	B11.17-1982(R1989)
Machinery and machine systems for processing coiled strip, sheet, and plate	B11.18-1985

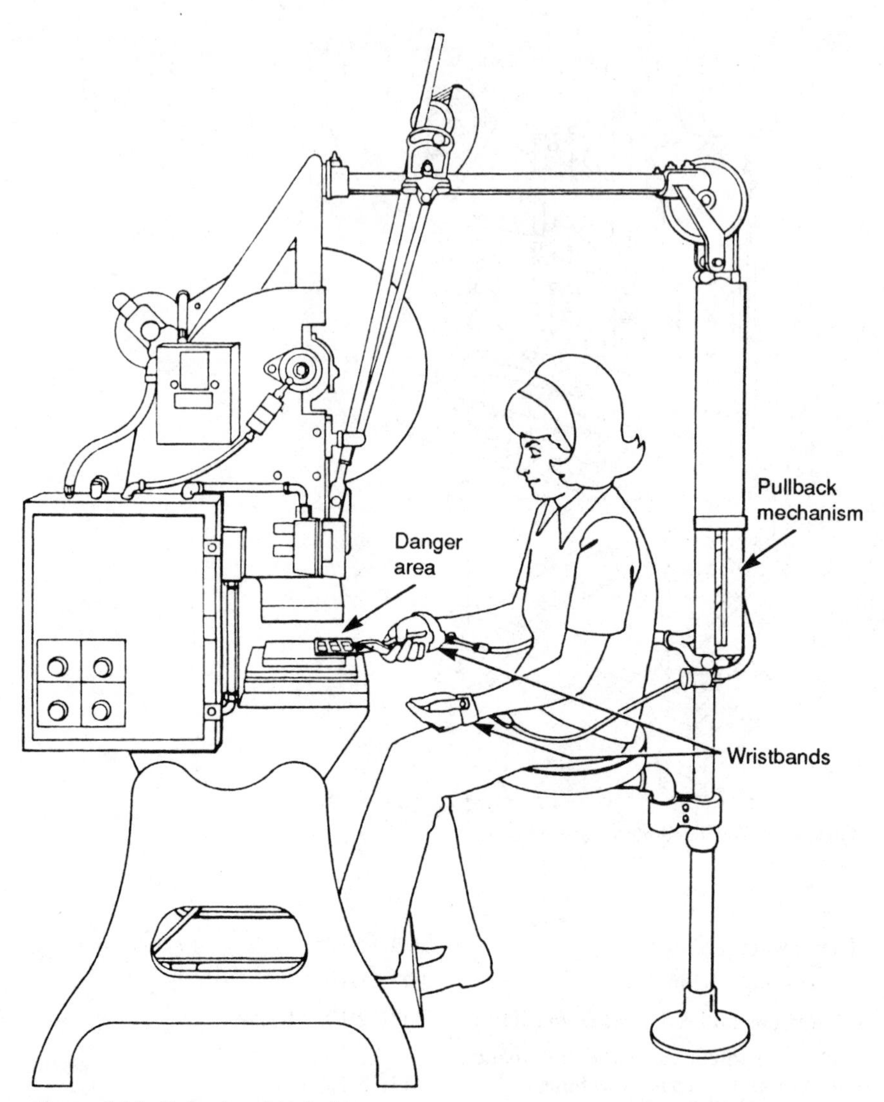

Figure 6.10. Pullback safety device.

Clearly there has been great progress in the development of safety standards for machine tools and similar articles in the past 20 years. As a matter of routine, designers of such equipment should begin with the most up-to-date standards as a reference point.

Forging equipment is regulated under OSHA section 1910.218, and its provisions are based on the 1971 edition of ANSI B24.1, *Safety Standard for Forging*. This standard is still carried under the same designa-

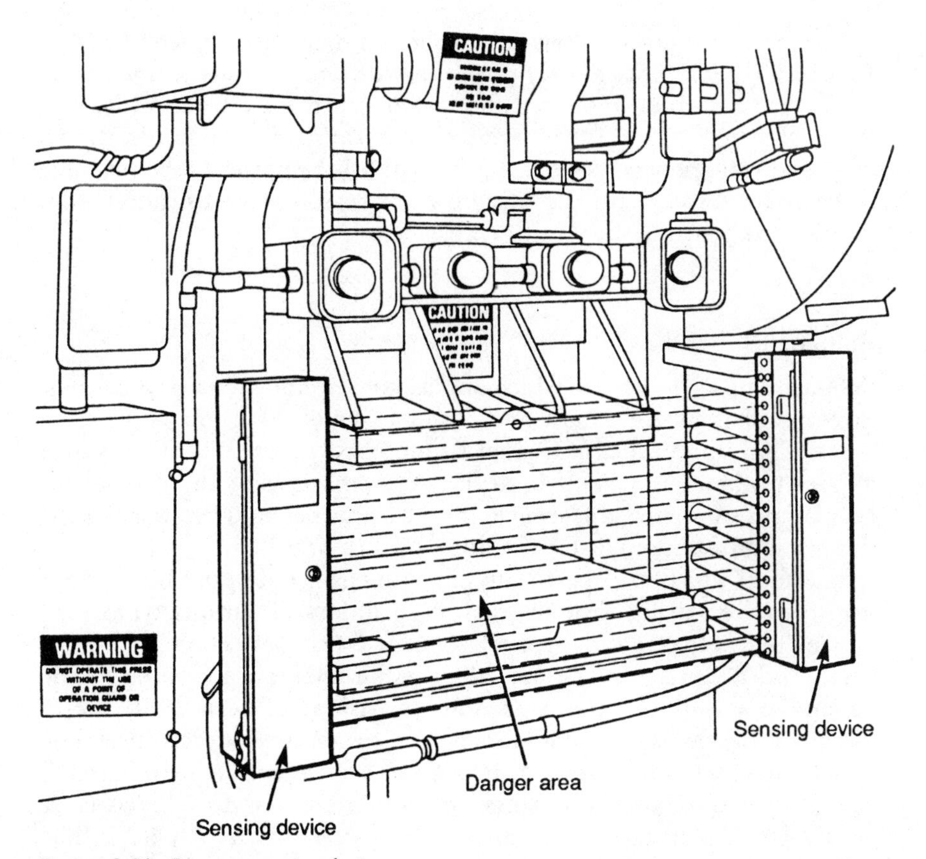

Figure 6.11. Presence-sensing device.

tion, but the most recent 1985 edition has been retitled *Safety Requirements for Forging Machinery.*

Design requirements for mechanical power transmission apparatus are set forth in section 1910.219. Most of the requirements relate to guarding for belts, pulleys, sprockets, chains, gears, and shafting as components of power transmission systems. Such systems were formerly used in factories which had a prime mover in the engine room. The prime mover, often a large reciprocating steam engine, developed the power which was distributed throughout the facility by the transmission system.

With the improvement in electric motors for individual machine drives, such mechanical power transmission systems have become obsolete. However, the belts, pulleys, sprockets, etc. which are components of the drivetrains of individual machines are still subject to the guarding requirements of OSHA regulations. The 1953 safety standard, on which the regulations are based, has been superseded by ANSI/ASME B15.1-1984, *Safety Standard for Mechanical Power Transmission Apparatus*. A typical floor-mounted fabricated guard for a belt and its pulleys is shown in Fig. 6.12. As a safety feature, many removable guards are fitted with interlock switches. They cut off the power to the machine drive motor if the guard is opened for any reason. An illustration of an interlocked guard is shown in Fig. 6.13.

Subpart P

Subpart P of section 1910, which applies to General Industry situations, gives regulations for Hand and Portable Powered Tools. The same types of products are also treated in subpart I of 29 CFR 1926, which applies to the construction industry. A comparison of the two sets of regulations shows few differences, but designers should be aware of the duplication of treatment for products of this kind.

The tools which are specifically mentioned include portable circular saws, powered drills, fastener drivers, grinders of various types, belt and platen sanders, and explosive-actuated fastening tools. For saws, drills, and fastener drivers, the safety requirements are covered by the general point-of-operation guarding provisions of section 1910.212, discussed previously. Grinders make use of abrasive wheels in a wide variety of shapes and sizes and may be either electrically or pneumatically driven. Safety requirements for them are covered by *Safety Code for the Use, Care, and Protection of Abrasive Wheels*, ANSI B7.1. The OSHA regulations are based on the 1970 edition of that standard, but the most recent edition was in 1988.

Belt and platen sanders have in-running nip points between their belts and the driving pulleys. As pointed out previously, these nip

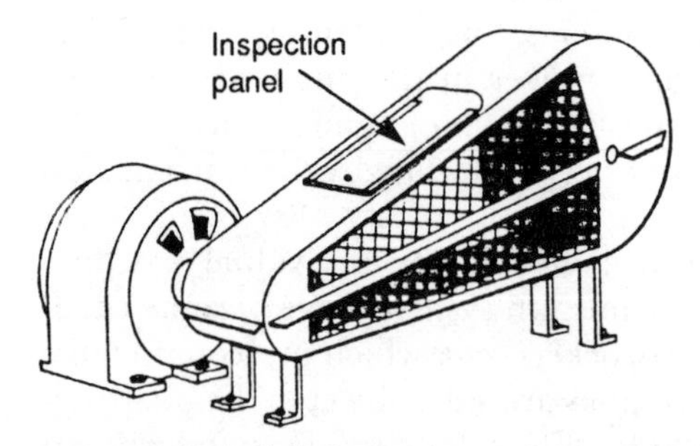

Figure 6.12. Belt and pulley guard.

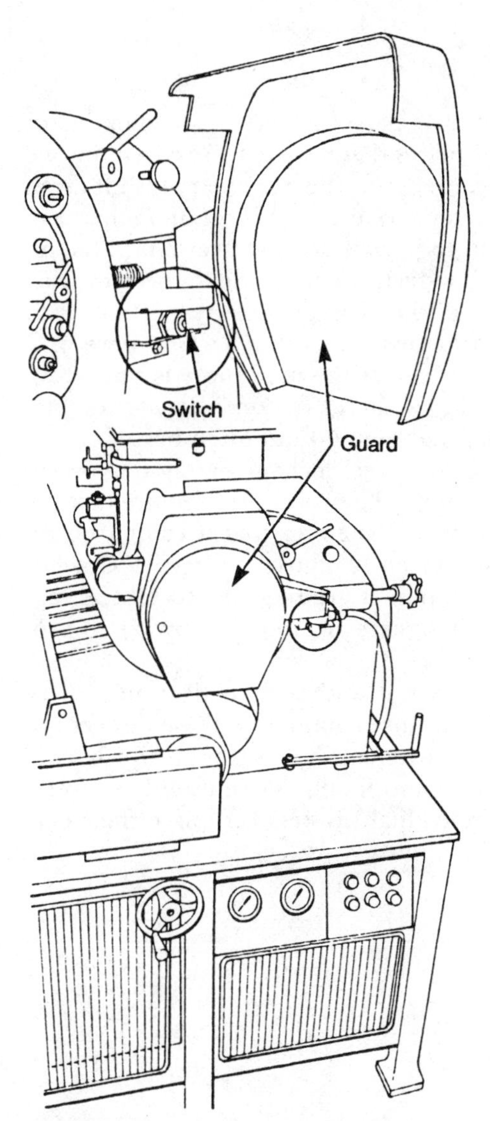

Figure 6.13. Interlocked machine guard.

points must be guarded. There are no published ANSI standards for tools of this type at present. The design requirements and safety considerations for explosive-actuated fastening tools are based on the 1970 edition of ANSI A10.3, *Safety Requirements for Powder Actuated Fastening Systems*. This standard was last revised in 1985 and should be used for design information.

Lockout and Tag-Out Provisions

All machinery and all equipment have parts that wear out, components that need periodic adjustment, or places that have to be serviced for one reason or another. When such events occur, the machinery is shut down for maintenance. During the maintenance effort, it is common that protective guards are often removed to permit access to certain parts, interlocks may be bypassed, or safety devices may be disconnected. During the shutdown, some members of the maintenance crew may have to put portions of their bodies into hazardous locations which were protected by the inoperative safety devices. If the machinery is started up again by someone who does not know that a worker is inside the machine, serious or even fatal injury may result—and often has.

To prevent such gruesome occurrences, lockout or tag-out procedures were developed many years ago. The procedures require that the worker secure the equipment in such a manner that it cannot be restarted by any other person than herself or himself. Effective October 31, 1989, all machinery and all equipment are required by OSHA regulations to provide for lockouts. Designers are required by law to provide this feature under 1910.147(c)(2)(iii).

Illustrations of lockouts for electrical and pneumatic controls are shown in Figs. 6.14 and 6.15. The feature common to these devices is a pair of mating tangs which have a matching hole through which a restraining device can be placed. The restraining device itself has several matching holes through which individual workers can place their personal locks to which they have the only key.

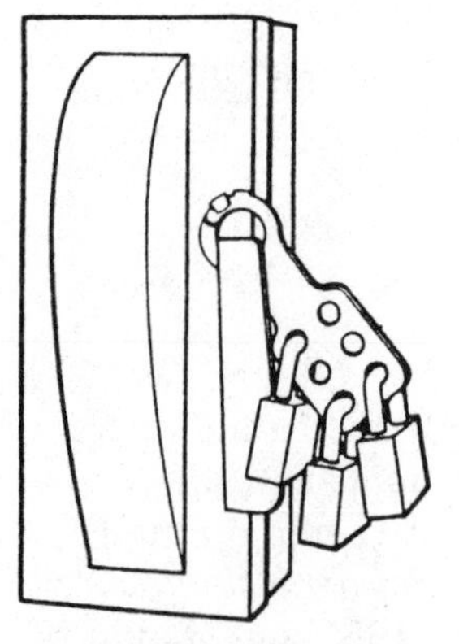

Figure 6.14. Electrical lockout.

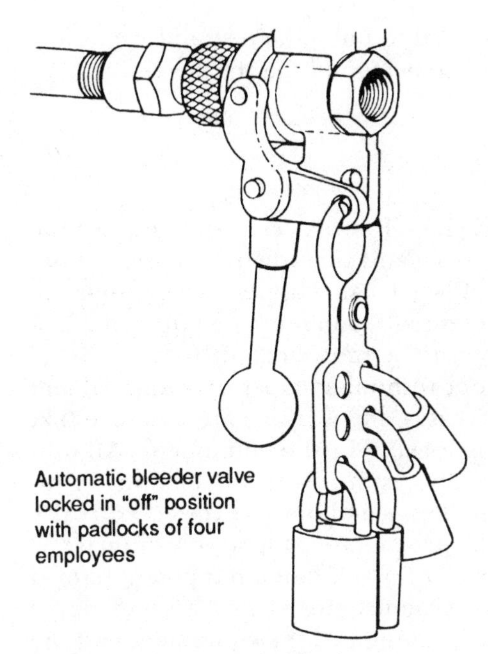

Figure 6.15. Pneumatic lockout device.

In practice, each worker who is going to be working on the machine during shutdown places his or her individual lock on the lockout device before entering the machine to begin work. The lock cannot be removed by anyone else until that worker has completed the assigned task and returned to the lockout location. In this way, all workers must be out of the machinery and clear of the hazards before the power can be turned back on.

Note that all classes of energy are subject to the lockout provisions. While electric power is the usual source for most machinery, all hydraulic, pneumatic, chemical, and thermal sources must also be controlled. In addition, the energy stored in springs and the potential energy of elevated weights must be taken into account. A thorough examination of all possible modes of inadvertent or unexpected energy release should be made during the design review activities, and suitable controls incorporated in the design.

Tag-out devices are similar to lockouts, but they involve the application of substantial and durable tags rather than locks. These devices are attached by a nonreusable link and sealed to prevent reuse. Such tags bear legends such as *Do not start, Do not operate,* and *Do not open.* The link is cut off the tag, and the tag is removed by the worker when the task is completed. This system is not quite as secure as the lockout method. No individual key is required to undo the restraint, and there is little protection for more than one person in the repair crew.

The Zero Energy State

How often we have heard the cliché, "But officer, I didn't know the gun was loaded." This is a time-worn and threadbare excuse, but obviously just another example of what can happen when someone plays around with a piece of potentially dangerous equipment. Unfortunately, the same scenario with a somewhat different cast of characters is frequently played out with many types of industrial and commercial equipment. In those cases, the workers are unaware that some lethal amount of energy is lurking in the equipment. All of a sudden it is released, and an injury results. Both the amount of released energy and its unexpectedness combine to produce surprise and shock, thereby increasing the probabilities for serious injuries or property damage. There is a remedy for this latent but potent hazard to life, limb, and property. First, recognize the hazard. Second, apply the zero energy concept to both the design of the equipment and the operating and maintenance instructions for it.

One simple method for designers to recognize the stored-energy hazard is to examine all the steps required to assemble the machinery and place it in service for the first time. During assembly, energy will be stored in every spring whose free dimensions are altered during assembly. Obviously, this applies to tension, compression, cantilever, or torsion configurations deliberately designed to store energy. Not so obviously, it also applies to cords, chains, and cables which may be under appreciable tensile loads. The loads cause the cords to stretch elastically, storing energy which can be released in a violent snapback.

In all cases involving springs, the designer must determine if the amount of energy stored is sufficient to cause danger. If it is, the circumstances under which the danger will manifest itself must be established. Then suitable lockouts, warnings, or alerting means must be devised. Their purpose is to avoid surprises by informing the equipment user of the nature of the hazard and how to avoid it. The user must be *told* when the gun is loaded. This is often done by installing lockouts or by placing placards on the equipment at the hazard location. In addition, it is common to make note of the hazard in the operating instructions.

Placing the equipment in operation will involve connecting it to a primary source of power, usually electricity, but perhaps some compressed gas or other fluid. As noted previously, OSHA now requires that lockout provisions for primary power sources be designed into all machinery and equipment.

When the equipment is first turned on after being connected to the primary power source, it is common for some components of the equipment to require charging before the machine will perform properly. This may be in the form of pumping up accumulators or surge tanks, charging actuator cylinders, loading hoppers with feed stock, energizing electronic power supplies, or lifting counterweights. All these actions result in energy of one form or another being stored in parts of the machine.

The zero energy concept requires that the equipment designer recognize that start-up operations can and do result in energy storage. It further obligates the designer to protect any operator or maintainer of the equipment from the consequences of unexpected release of that energy. The analytical key to this design problem is the preparation of an energy audit. The audit is based on the designer's knowledge of all the places where energy is to be stored or used in the product. It provides the clues to what needs to be protected, where it is located, how much energy is involved, and how the hazard can be eliminated. It also provides guidelines for the preparation of start-up and shutdown instructions for the equipment.

Instructions for shutting down equipment should begin with procedures for turning all operating controls to the off position. Next, all energy sources to the equipment must be disconnected or locked out. Accumulators must be bled down to atmospheric pressure. Electrical capacitors must be short-circuited to discharge them. Elevated weights must be blocked up or lowered to their rest positions. All loose or freely movable components of the equipment must be secured to prevent their falling out during subsequent work periods. All rotating parts, particularly flywheels, must be brought to a complete stop. For complicated equipment, it may be necessary to prepare and follow a checklist to ensure that no source of energy is overlooked. Input from the equipment designer is mandatory during the preparation of such a checklist.

When the shutdown and lockout have been completed, make certain that all personnel are clear of the equipment, then operate the various controls to test and verify that the machine is actually in the zero energy state. Make absolutely certain that all such controls are returned to their off positions after this test is completed.

During some maintenance procedures it sometimes becomes necessary to try out the equipment to check the proper performance of certain features. If the performance is not acceptable, additional work is required. In such instances, the entire shutdown and lockout procedure must be observed again.

Examples of Commercial and Industrial Product Designs

At the beginning of this chapter, we said that one characteristic of commercial and industrial products which distinguishes them from consumer products was their relatively long service life. From the economic point of view, long service life is an advantage. However, from the safety end of the telescope, long service life is a different story. It serves to perpetuate design problems for years after safer designs have been developed and put on the market.

The National Safety Council estimates that well-guarded machinery, with the guards designed in accordance with the recent standards and built into the machine by the manufacturer, accounts for less than 20 percent of all machines currently in use. Of the remaining 80 percent which were built to older requirements or used add-on guarding, the NSC estimates that less than 20 percent have been brought up to current standards of hazard mitigation. That NSC estimate indicates that about two-thirds of all machinery in use in the United States today has one or more safety defects. These defects had their origin in the design. If the hazards had been recognized in the design stage, they could have been designed out of the product, as proved by recent models. Instead, long service life extends the duration of the hazardous conditions.

Mechanical Power Presses

Punch and stamping presses are outstanding examples of this situation. With little more than routine maintenance, they last for decades. Often they are used for a few years by one owner, then sold and sold again to other operators. None of the owners bothers to bring the press up to current safety standards. The point of operation is the critical location for this equipment. This is where the dies and the workpiece come together when the press makes its working stroke. Too many workers have had their fingers or hands caught and mangled in power presses which were improperly guarded. This carnage goes on, even though designs for effective guards have existed for years.

Unfortunately, current design standards assign the responsibility for point-of-operation guarding to the press operators. The rationale is that each press application is unique and requires a custom-designed guard. The press manufacturer is in no position to know what application the operator will use the press for and therefore cannot design the guard for the operator. No universally accepted point-of-operation guard has been designed for mechanical power presses yet.

One attempt at a universal point-of-operation guard was introduced several years ago. It consisted of a swinging stick which was attached to the ram of the press through a hinged linkage. When the ram came down, the linkage caused the stick to sweep across the point of operation of the press. Theoretically, the worker's hands were swept out of the danger zone. Actually, if workers failed to get their hands out of the way of the stick quickly enough, they could get a broken wrist. Because the guard itself had the potential for inflicting injury, such swing-type guards were outlawed.

The author has investigated several personal injury cases which involved improper guarding for power presses. In one instance, the press was a small vertical-stroke machine which was used to punch holes in sheets of paper or cardboard for office file folders. The press was driven by a ¼-horsepower electric motor, which was belted to the flywheel of the press and drove it at a relatively high speed—so fast that when the operator pressed the foot pedal to actuate the press, the ram came down, punched the required holes, and returned to the start position in less than half a second. Clearly, if the operator had any fingers in the danger zone at the start of the stroke, there was no time to get them out of the way of the moving ram.

The guarding, such as it was, consisted of a piece of clear plastic in front of the point of operation and extending about 8 inches to either side. Both ends of the point of operation were left open. The work was put into the point of operation by sliding it underneath the plastic guard. The operator could see through the plastic to check that the workpiece was in the proper position before stroking the press. If something got stuck in the press, the operator could reach around the end of the plastic guard to remove the obstruction. Sure enough, something did get stuck in the press, and the operator did reach around the unprotected end of the guard, and just then the worker accidentally pressed the foot control. The result was two lost fingertips. The reason was inadequate guarding. The guard did not make it physically impossible for the operator to get any body part into the danger zone during a press stroke. This is a clear violation of the requirements of OSHA regulations, 29 CFR 1910.212. It is also a totally correctable design error by the person who made the guard.

Press brakes are a variation of power presses. They are used to form sheet metal by a bending operation and are distinguished by having a very wide lateral opening as the point of operation. The width of the opening makes it possible for two persons to work side by side at the same time. Of course, this also means that the workers coordinate with each other in the operation of the controls. The foot pedal control to actuate the machine stroke consisted of a long horizontal bar which extended nearly across the entire throat width of the brake. It was therefore possible to stroke the press from any point across its width. The point of operation was completely open from all sides. No pullbacks were provided for the operator's hands, nor was there any proximity-sensing device in operation across the point of operation. Again, this is a clear violation of the machine guarding regulations, one easily correctable with available devices.

The scenario in this instance was that two students in a metal shop class were given an assignment to bend sheet-metal panels to make some shelves. This was a two-person job. They worked in tandem, standing alongside each other. One student bent the long edges of the sheet, then passed the piece to his partner to bend the short sides. When completed, the finished pieces were stacked on the floor. After they completed one batch, the first man leaned down to pick up some more sheet stock. In so doing, he placed his right hand in the point of operation to steady himself, forgetting that his partner still had a couple of strokes to make to complete the last piece. The brake crushed three fingers.

Another stamping press investigation involved a worker who reached inside a seemingly well-designed point-of-operation guard with disastrous results. This press was set up to work on an automatic cycle. It ran unattended, making continuous strokes as strip sheet stock was unwound from a reel and fed through a slot into the point of operation from the right-hand side. Finished parts were ejected out the rear, and the scrap stock was pulled through a slot on the left side by a take-up reel. The left-hand slot was part of a hinged door provided to allow access to the point of operation in case something got jammed up. The key feature of this door was that it was not interlocked with the control circuit for the press. Another factor which contributed to the causation of this accident was that the worker was not given any retrieval tools to clear the jam.

When the machine jammed, sensors stopped the automatic cycling, and the worker went over to clear the machine. He did not shut off the power, but opened the door in the left-hand side of the cage-type guard surrounding the point of operation. Without a retrieval tool to use, he reached in with his bare hand, cleared the jam, and lost two fingers when the machine began to cycle itself again. This loss was easily preventable in spite of the worker's failure to shut off the power to the press. An interlock on the door of the guard would have prevented the press from stroking while the door was open.

Food Preparation Equipment

While food processors are relatively new to U.S. households, they have been used in restaurants and commercial food plants for many years.

One of the more common machines is used to chop food. A slowly rotating tray passes the food in a clockwise direction under a set of rapidly rotating, very sharp knives. The knives and most of the tray are covered by a heavy cast-metal hood. The hood has a hinge along the rear edge so it can be swung up and out of the way for cleaning or for clearing out a blockage. An interlock switch cuts off the power to the electric motor when the hood is lifted. The hood has an opening in its top surface through which the food is fed to the tray. From the safety standpoint, a key feature of this design is the size of the feed opening. In its early design version, the opening spanned a pie-shaped sector from about the 5 o'clock to the 9 o'clock position. Another key feature was the shape of the rotating tray. It had a semicircular groove about 4 inches wide running around its circumference. There was a matching groove in the bottom side of the hood to provide a doughnut-shaped path for the food to move around and under the cutting blades. Since it was 4 inches wide, it was large enough that someone could insert the left hand and fingers into the groove without having to raise the lid.

Foreseeably, one day a kitchen worker had trouble with the machine when it got blocked by some chunks of food. Instead of raising the lid, which would have cut off the power, the worker stuck his hand into the groove and tried to pull the blockage out backward. The tips of three of his fingers were cut off when he touched the whirling blades.

Obviously the worker made a gross mistake when he reached into the moving parts of the machine instead of opening the lid. However, it is just this sort of blunder which must be anticipated by the designer. Even rudimentary testing of this machine during its development stages would have shown that such an accident was possible and that there was an unprotected hazard which was easily recognizable. Furthermore, the hazard could be designed out of the product simply by making the opening smaller. Later versions of the machine incorporated that obvious design improvement.

Band saws, very similar to those used for many years to cut wood in lumberyards, are now becoming common in butcher shops. Using a special blade, a meat cutter can saw pork loins into pork chops very efficiently. By using a preset gauge, the chops can be turned out uniformly to any desired thickness. Because of the sanitary requirements applicable to food preparation equipment, several special design features not found in the woodworking version are included to make the equipment easy to clean. Some meat-cutting saws also have an inclined blade which provides a modest self-feeding action. This makes it easier for the operator to move the meat past the blade, increasing productivity and decreasing fatigue. It also makes it easier for the operators to cut themselves.

There are other significant differences. In the wood-cutting version, the operator stands facing the toothed edge of the blade and pushes the workpiece forward on a stationary table. Meat cutters place the workpiece on a table which slides back and forth sideways. They stand facing the side of the blade, moving the table back and forth to make successive cuts in rapid sequence and feeding the workpiece laterally into the blade one slice at a time. As long as the piece of meat is sufficiently large, their fingers are far removed from the danger zone at the point of operation.

The problem arises when the last few cuts are to be made on the piece of meat. Then the operators get their fingers close to the hazard of the moving blade. If there is a hang-up or the blade wobbles out of line, contact with the exposed portion of the tooth-carrying side of the blade is probable. One obvious solution is to use a push piece. These are supplied with the saws when they are first sold, and a separate rack for storing them is designed into the movable carriage. Warnings are prominently displayed on the machine covers, and some operators wear protective finger guards. But still, meat cutters continue to be injured. What is really needed is a blade guard which will be effective but not degrade the productivity of the machine. There is a real challenge for designers.

Dough brakes are small rolling mills which bakers use to roll out chunks of dough. They do the job of the old-fashioned wooden rolling pin, but they do the job mechanically and much faster than before. To use a brake, the baker places a chunk of dough on a heavy metal table located in front of a set of three powered rollers. The dough chunk is then pushed manually into an adjustable gap between the first two rollers. These rollers are located one above the other. The third roller is behind the other two and is used to return the dough for additional working.

The key hazard with this equipment is the in-running nip point between the first pair of rollers. If the bakers push too far forward on the piece of dough, their fingers or hands will be drawn into the nip along with it. This hazard was recognized many years ago. It was designed out of the machines in two ways. The first was to make the table long enough to prevent the operator from being able to reach into the nip point from the normal working position. This approach is known as safety by separation or location.

Just to make absolutely sure, a second protective feature was designed into the machine in the form of a pressure bar along the rear edge of the table. If the operator leans too far forward, any body contact pressure on the bar shuts off the power to the drive motor and actuates a powerful brake which brings the rollers to an instantaneous stop. This design was conceived long before OSHA came on the scene, but it was so successful that its features were incorporated in the regulations when they were published in 1971.

Cranes

One of the more challenging investigations involved the design of a cherry picker crane. It was being used by a utility line worker when the boom suddenly collapsed and dropped the bucket to the ground with the line worker still in it. The position of the boom was controlled by a set of hydraulic cylinders which could be actuated from a valve panel in the bucket. Examination of the cylinders showed that one had apparently failed under an excessive tensile load. This was most astonishing since these cylinders were never intended to carry tensile forces, only compressive loads.

Several theories of material failure were invoked, but none could account for the observed facts and the condition of the failed cylinder. It was thought by some investigators that the cylinders had been damaged by accidental overpressuring when they were tested prior to installation. This did not account for the obvious pulling off of the large nut on the end of the piston rod of the failed cylinder.

The successful approach to finding the cause of the failure lay in backing into the problem to determine the circumstances required to produce tensile loads on the cylinder. A study of the kinematics of the boom layout showed that this could happen only if the boom were driven past a certain critical point where the boom and bucket were overcentered. In that position the whole boom became unstable, and tensile forces developed in the piston rod.

The primary question became, was it possible for the line worker's controls to drive the boom into the overcentered position? The answer was found by examining the limits to motion on the kinematic diagram. It immediately became apparent that there had never been a limit stop switch installed to prevent the boom from being driven over center. There was no mechanical stop installed either. Obviously no one had even done a stability analysis of this product, nor had it been tested for stability before being put into production. It is a sad commentary that a line worker had to suffer crippling injuries before these obvious design gaffes were corrected.

Still another mystifying series of failures took place in a fleet of small mobile crane trucks. They were used by a telecommunications utility to service wire and cable installations which extended over a sizable area. They logged many miles of both on- and off-road travel every month, in addition to their regular use in positioning a line worker's bucket at a static location.

One day the line worker was driving the crane truck along a rough road when he heard a strange banging noise coming from the rear of the truck. He stopped the truck, got out, and looked at the crane bucket in amazement. The knuckle joint which secured the bucket to the boom had failed—failed in its stowed position while it was not even being used! The failure allowed the bucket to twist to one side, and it would probably have dropped the line worker to the ground if it had happened in service. And this truck was only a few months old.

Investigation of the failure focused on the types of forces and torques acting on the knuckle joint under various uses. The resulting stresses were calculated and found to be conservative. However, optical examination of the fracture site showed clearly that the failure had been a progressive one. A crack had originated at an inside corner of the joint and grown across the throat of the knuckle until there was not enough sound material remaining to carry the applied load.

But what was the applied load if the stress calculations showed such conservative values? Because of the progressive nature of the failure, metal fatigue was obviously involved. This implied that the load had to have been applied many thousands of times. But in the short service life of this unit, the line workers had not applied their working loads more than a few hundred times. There was no way that the intended use of this equipment could have resulted in this type of gross mechanical failure.

That conclusion eliminated the infrequently applied normal working loads as the probable cause of the failure. However, it left the investigator with the problem of figuring out what other loads must have been present, applied frequently, and big enough to cause fatigue damage. A review of the driver's report provided the necessary clue. The driver had been going over a rough road when the failure occurred. Of course, the truck bounces when it goes over bumps. Bingo—there is the frequent event. But did it cause large loads on the knuckle joint? Yes, because the boom was not tied down when it was in the lowered position. The designer forgot that the cantilevered boom could act as a spring and actually amplify the normal dynamic loads on the bucket. The designer also forgot that stresses on the inside corner of the knuckle joint could be several times their nominal value because of the presence of a stress concentration factor at that location. Put those two design errors together, and the result is absolutely predictable. The only question is when, not if.

Lockouts

Mishaps in which a worker got caught in a screw conveyor were among the most depressing cases in the author's forensic engineering career. They were depressing for two reasons. First, they could have been avoided so easily; second, they always resulted in gruesome injuries to the victim.

In every case, workers were doing maintenance work on the conveyors. They were working at a location which could not be seen from the electrical panel which controlled the power supply to the equipment. In every case, another worker turned on the power without checking to see why the power was off. The reason became apparent immediately.

These accidents occur all over the United States, and involve many types of industrial operations. The common factor is the moving of material by means of a helical screw which fits closely inside a circular or semicircular channel. The clearance between the inside wall of the stationary channel and the outer edge of the rotating helix is purposely made small, to prevent material from leaking backward along the channel. It is a geometrically perfect nip point, and the edge of the moving helix acts just as a guillotine on anything that gets caught in there. Amputations of fingers, hands, or feet are the common result.

Designers can eliminate this recognizable hazard by two methods. The first method is to specify lockout tangs on the switches which control power to the conveyor drive motors and to provide matching hasps to be used by the workers during shutdowns. The workers are expected to provide their own individual locks and keys for securing the hasps to the lockouts. Management is expected to have and enforce an effective lockout procedure. The second method is to identify all points on the equipment which will need periodic maintenance and to design removable guards for those locations. The guard should include an electrical interlock feature which will automatically cut off the power when the guard is removed, even if someone forgets to lock out the main power switch.

Special Machinery

Did you ever wonder how they wash the windows on one of those new high-rise buildings, especially the ones with the windows that cannot be opened from the inside? They do it with a mobile scaffold that moves around the roof on a set of rails and carries a platform which is suspended from a set of cables. The window washers and their equipment are moved up and down on the scaffold by a set of electric winches. The winches are synchronized to keep the working platform level at all times and are under the control of the workers on the platform.

Because of the potentially spectacular consequences of any kind of mechanical failure in equipment of this sort, designers use very generous safety factors, specify the highest-quality materials, and incorporate multiple duplicate safety features throughout the entire power train.

In spite of all this cautious engineering, there have been instances of power train failure. One which was investigated by the author involved the failure of a drive shaft in one of the pair of winches. The winding drum of the winch let out several feet of cable before the overspeed brake stopped it. The sudden drop and subsequent sudden stop caused the worker on the fallen end to lose his footing. Unfortunately, the platform was just starting its descent from the roof, and that window washer had not fastened his safety belt to the scaffold railing. He fell more than 50 stories, landing on a paved walkway. The other worker, who had secured his safety belt, dangled from the platform until he was rescued by a courageous firefighter.

Examination of the failed drive shaft found no defects in the material itself. All its physical properties were well within specified limits, and the part had been machined in accordance with the applicable drawing. It was concluded that neither the material selected nor the manufacturing method used was responsible for the failure.

The appearance of the surface where the shaft broke and the location of the break revealed the reason for the failure. The break was across a section where there was a significant change in the diameter of the shaft. The corner between the two different diameters had a very small radius, which acted to concentrate stresses there. Significantly, there was no call-out for that radius on the drawing. The omission of that detail by the designer cost a window washer his life.

Human factors played an important role in a mishap involving the rollover of a vibratory pavement compactor. Ordinarily, equipment of this type is so heavy and so stable and moves so slowly that the possibility of its turning over is simply ignored. This, however, was no ordinary circumstance when a young construction worker was given brief operating instructions by his boss and then told to drive the compactor up a hill to the work site.

Two design features of this equipment made it unusual. The first was that it consisted of two separate powered rollers mounted on a strong frame that was hinged in the middle to enable the unit to be steered. Hydraulic cylinders swung the hinged frame in accordance with a servo output from the steering wheel. The second unusual design feature was the use of a hydrostatic transmission in the power train to the rollers.

There is nothing esoteric about hydrostatic transmissions. They have been used successfully for many years on myriad applications. They are very simple to operate. All you do to make the unit travel forward is to push a control lever forward. The farther you push the lever, the faster the unit will go. There are no gears to shift, no clutches to engage. You do not even have to work the throttle since the engine runs at constant speed. To make the unit go backward, pull the lever back. The farther, the faster, as for the forward direction.

This device is simple, yes, but there is one peculiarity which can be confusing to a poorly trained operator such as the victim in this turnover accident. The peculiarity is in the way the brakes are applied when the operator wants to stop the roller. There was a foot-operated brake on this unit, but it was small and used only to hold the roller in park after the engine was turned off. The real way to put the brakes on this machine was to put the drive control lever in *neutral.* Yes, neutral. In the neutral position the transmission is locked up hydraulically. The rollers cannot be forced to move as long as the engine is running to drive the hydraulic pump which transmits the power.

This is in total contrast to the way that a hydrodynamic transmission works. Hydrodynamic transmissions are the automatic type usually found on passenger cars and some trucks. The neutral position on the shift lever decouples the engine from the drivetrain and lets the vehicle roll freely. The young driver of the compactor was familiar with hydrodynamic transmissions. He was conditioned by his previous experience to expect that neutral on the control lever would allow the vehicle to move freely rather than coming to a stop.

This human factor came into play when the compactor began to slip on a grassy part of the hillside. The driver never thought to put the control lever in neutral to stop it, but tried to steer his way out. He lost control when he turned too sharply on the side of the hill. Fortunately the flip-over threw him clear of the machine, but it was heavily damaged.

A woodworking machine known as an edger poses a challenging design problem. These machines are used in sawmills as the next operation following the sawing of a slab of wood from the bole of a harvested tree. As the cut slab comes down from the big buzz saw, it will have flat sides but the edges will be uneven because the tree trunk is never perfectly straight. To true up the edges, the slab is run through another machine which has two or more saw blades. These blades can be placed various distances apart in order to make boards of different widths.

During the edging operation, it is important to prevent wandering by holding the workpiece down securely and to feed the wood at an even speed. A set of hold-down rolls provides this action. When the squared-up boards are driven out of the edger, they slide onto a set of non-powered rollers which are part of a receiving table. A worker on the output end of the machine pulls the scrap material off the table, puts it in a chute, and feeds the good boards to the next saw where they are cut to length.

The design problem comes up if a workpiece has a thin portion at its rearward end. This condition is known to lumbermen as *wane.* It occurs every once in a while, but often enough to be a nuisance. When the back end of a board with wane passes under the pressure rolls, the change in thickness forces the back end of the board to move downward under the roll pressure. Downward movement of the rear end of the board automatically causes the front end of the board to fly upward. Often, there is another board in contact with that board. The second board can be thrown upward or to one side with considerable violence. The worker at the output end of an edging machine is completely exposed to this severe hazard which needs to be designed out of this equipment.

Summary

Industrial and commercial machinery and equipment are intended to be used by employees at the work site. As a result, the designs of such products are subject to the regulations of the Occupational Safety and Health Act of 1971. By law, employers must provide a workplace which is free from recognized hazards. This requirement implies that employers cannot legally operate equipment which presents unprotected hazards to workers. In turn, designers of such equipment are obligated to design such hazards out of their products before they are placed on the market. As a result of OSHA, no sane employer would purchase machinery or equipment which does not meet the OSHA safety requirements.

Any aspect of the equipment design which relates to safety falls under the purview of the OSHA regulations. So designers must make a survey of applicable OSHA regulations very early in the concept phase of the design. Many OSHA regulations are based on standards which are seriously out of date. Only the latest versions of applicable standards should be used for reference. The current catalog for ANSI or other pertinent standards-writing organizations should be consulted for the most up-to-date information.

Design reviews for commercial and industrial products should include both hazard recognition efforts and an energy audit. The energy audit is intended to provide guidance for the design of lockout and tagout procedures, as required by the 1989 revision of the OSHA regulations. The hazard recognition effort gives designers the opportunity to design the recognized hazard out of the product at once. If the hazard cannot be designed out of the product, guards must be incorporated into the existing design. If no suitable guards can be devised, effective warnings and instructions must be prepared.

7
Codes and Standards for Design

Although the design process is essentially an exercise in creative problem solving, many previous attempts on record can often guide designers in formulating the approach to a specific problem. In the mechanical design areas, many of these design solutions have become commonplace and have been collected into what is generally known as *good practices*. When used by many separate and distinct designers, the solutions become recognized as the accepted way to do a certain thing. After this approach has become accepted, it may be formalized into a standard.

Use of the standard may be voluntary in some instances, but mandatory in others. In all cases, however, designers are well advised to become acquainted with any existing standards which apply to their particular problems. This survey of applicable standards should be one of the earliest tasks in a design project, following closely behind problem definition and the establishment of an approach to its solution.

History

Engineering students are all familiar with the fundamental physical units of mass, length, and time. All physics texts contain descriptions of the standard kilogram and the standard meter bar, which serve as the bases for the metric system. The scientific community recognized, long ago, that such standardization was necessary for the valid replication of observations in different laboratories by different investigators.

The meter was intended to be one ten-millionth of the distance from the equator to one of the earth's poles when it was established in 1790, but later measurements showed that this was not quite accurate. From 1889 to 1960 the meter was defined as the distance between two lines scribed on a platinum-iridium bar which was kept at the International Bureau of Weights and Measures in France. In 1960 improvements in metrology resulted in a change in the definition to one based on the wavelength of a certain orange-red line in the spectrum of krypton 86. Then 23 years later, the definition was changed again, and now it is based on the speed of light in a vacuum. It is clear, therefore, that even among the most basic physical quantities, *standard* does not mean something that is fixed and unchangeable for all eternity. Periodic review of published standards is a never-ending process.

While the need for standards was obvious to the scientific community and was accepted by those practitioners, it was not so well recognized by the early applied scientists and engineers. They continued to do things in their own individual ways, and so each design was unique. There was no interchangeability of component parts from one item to another. Everything was almost custom-made.

Industrial standardization was first conceived in the United States by Eli Whitney. Although remembered by most people for his invention of the cotton gin, Whitney never received much financial reward for that noteworthy contribution since it was easily copied by any skilled mechanic. Whitney's real contribution was the invention of the jig. The need for this invention arose from a government requirement that parts for certain rifles be sufficiently similar that they could be exchanged with each other and still have a working rifle. This is recognized today as the requirement for *interchangeability*, but it required the invention of Whitney's jig to permit individual workers to control the dimensions of successive parts within tolerable limits. The high-volume production characteristic of current manufacturing would be impossible without standardization of dimensional tolerances.

Interchangeability between parts made by a single manufacturer is only one side of the standardization problem, however. A more complex situation arises when attempts are made to match up parts which are made by different companies, each according to their individual designs. This is the problem of compatibility. In many examples lack of compatibility created serious problems. Among the most egregious was the use of different hose and hydrant threads by various fire equipment manufacturers. The readily foreseeable result was that the fire companies from one town could not assist those in an adjoining town in the event of a major emergency. Because of the lack of compatibility between the different threads, hoses from one town could not be connected to the hydrants of the other town. Even though all the different thread types worked well in service, the lack of compatibility eventually forced the standardization of hose threads throughout the United States.

Compatibility shows up in many other ways. The determination of which side of the road vehicles are to be driven on is an arbitrary decision and is not universal. Great Britain and Japan require drivers to use the left side of the road, while most other countries drive on the right. Some Scandinavian countries changed sides a few years ago, so the choice is not always an eternal one. Railroads are similarly bound by a choice of direction. The Chicago and Northwestern Railroad was built by British engineers and used the left-hand system, while the U.S. roads followed the right-hand system. Even the gauge of railway tracks is not universally the same. Railroad cars traveling from Europe into the Soviet Union must change to different sets of wheels and axles to accommodate the wider track gauge used by the USSR. This lack of compatibility is expensive and time-consuming, but changing the gauges now would be prohibitively expensive. Thus some choices of standards become literally irreversible. Standards writers bear a heavy responsibility because of the relative permanence of what are often rather arbitrary choices.

The Standards-Writing Process

Once the technology for a certain product or process has matured to the point where *good practice* can be defined and the proliferation of unique solutions to the design problem has diminished, then the situation is ripe for the preparation of a standard. The usual procedure is for the various parties interested in this particular product or process to get together in some formal or informal way and agree among themselves that it is necessary and desirable to prepare a standard. Once this agreement is reached, the work can begin.

The circumstances which call for the preparation of a standard are many and varied. In the simplest cases, the need arises within one company and is often brought about by the requirement for uniformity of its products. To meet the uniformity specifications, methods must be developed to control certain of the production processes. The people involved in the process get together to work out a method of achieving the required goal, and they formalize their findings with an internal process standard for use throughout that particular company. Commonly such standards are required more than once, and they eventually are collected into a manual of standards. This is issued by the technical department for use by purchasing, manufacturing, and other departments within the company.

The format for these *company standards* will vary from one organization to another. The group which actually writes the standard normally includes members from all the departments involved. All members in turn get to put in their two cents worth, reflecting their concerns on how the implementation of the proposed standard will affect their operation. Several drafts of the standard may be prepared before general agreement is reached. This is the *consensus* process working at the local level. However, it is used at the level of national standards as well.

The next step up from company standards is the development of *industry* or *trade* standards. In this context, representatives from several organizations which are engaged in similar businesses or which make similar products get together and establish a consensus on the details or features of their products. The intent here is usually to ensure the compatibility of products among various makers. Examples abound. Early color TV sets used two different systems, but only one survives today. Videotapes still are made in two different formats, VHS and Beta. There is a large effort under way to formulate a common method for computers to communicate with each other. Bumper heights for automobiles are all about the same, to avoid overriding between different cars.

The automobile industry offers a classic example of industry operation on standards. Efforts by automobile manufacturers have led to the establishment of a separate engineering society devoted primarily to automotive design problems. This, of course, is the Society of Automotive Engineers, known as the SAE. There are literally hundreds of standards published by SAE relating to self-propelled vehicles. Other countries have similar organizations.

Although the intent of industry standards is usually the benign one of ensuring compatibility among the products of various manufacturers, the U.S. government has taken a hard look at the possibility of collusion in the carrying out of the consensus process. In the spring of 1990 hearings were held on this subject to determine whether standards have any anticompetitive effect which tends to squeeze out certain organizations in favor of others. It appears that one person's consensus may be another person's restraint of trade, hence illegal in the United States.

As an industry matures and expands, commonly its standards are promulgated on a national basis. The usual avenue is through the American National Standards Institute, known as ANSI. The sponsoring trade association will submit a request for consideration of its standard by ANSI. A review group is then formed which includes members of many other groups besides the industry itself. This is an effort to provide a broader consensus for the national standards than that provided by the industry alone. Proposed drafts of the standard are circulated to all interested parties for comment, according to a long-established procedure. Notices are published at regular intervals, announcing the availability of such draft standards and inviting comments. A time frame is set up for receipt of comments, after which the Board of Standards Review considers the comments and makes any changes which appear necessary in the proposed standard. After several reviews, the standard is finally issued and published by ANSI. The standard is then listed in the ANSI catalog and is available to anyone who wishes a copy for the price listed.

The unrestricted availability of the published standards is one of the salient features of the present system. Whereas company or industry standards may not be accessible to persons or firms outside the particular company or industry, published national standards are available to everyone. Many technical libraries have excellent collections of published standards and make them available for reference purposes. With this wide dissemination of information and its easy accessibility, there is no valid reason or excuse for design engineers to fail to review available standards in the early stages of their design efforts.

Another very large standards-publishing organization in addition to ANSI is the *American Society for Testing and Materials,* known as ASTM. As their name implies, the scope of their efforts is directed primarily toward the establishment of uniform methods for testing various properties of materials and the definition of those properties. ASTM works on a committee system, and there are hundreds devoted to various classes of materials and testing methods. ASTM uses the consensus method of reaching agreement on what the standard should be, and there are several reviews, round-robin testing procedures, and comment periods before the standard is first issued as a tentative specification. A large capital T is appended to the designation of any new standard to indicate that it is still tentative and subject to revision before being issued in final form. ASTM has a catalog of its publications, but most of its standards are collected in bound volumes which combine the standards relating to a specific class of materials. The entire set of standards, which now numbers in the tens of thousands, fills a large shelf. Once a person finds the applicable ASTM standard, however, it is possible to obtain a single copy of the standard for a fee. As for the ANSI standards, many technical libraries maintain a set of volumes of ASTM standards for reference purposes.

Another major developer of product standards is *Underwriters Laboratories, Inc.,* also known as UL. The impetus for UL was the advent of many uses for electrical devices in the early part of the twentieth century. Some devices presented severe fire hazards which caused losses to the fire insurance underwriters. In an effort to reduce these losses, the insurance companies set up UL as a testing and certifying agency. The staff members of UL draw up a standard for testing of a certain device entirely within their own facility. The UL standards are not written from a consensus basis, in contrast to the ANSI and ASTM standards. The UL standards are intended to reduce insurance losses by imposing certain performance requirements on the products to be tested. In most cases these performance requirements leave it up to the designers to figure out a way to meet them rather than prescribing a certain specification for the design.

If a product meets the UL performance test requirements, it is eligible for certification of that fact. This may be done by attaching a small label which carries the UL logo or by other suitable marking. At periodic intervals, UL publishes lists of approved devices which can be used for reference by designers. Significantly, many products are tested on a repetitive basis to ensure that the performance shown in the original tests is maintained by the manufacturer. This is known as *reexamination service* by UL. In addition, UL field inspectors make periodic visits to manufacturing facilities to verify that product integrity is being maintained.

Copies of UL standards are available upon request and are listed in a catalog, but are not as widely available in libraries as the ANSI and ASTM standards are. The UL standards are usually not collected in volumes, but are used singly for specific applications. As with the other standards, designers should review UL standards in the early stages of any design effort. In many instances, UL certification may be a requirement for selling a product in certain jurisdictions.

Government, Industry, and Commercial Standards

Government Standards

In many countries, and particularly in the United States, the national government is one of the largest purchasers of products of all kinds. It is, therefore, a major market to be served by many different manufacturers. To ensure that the product to be supplied to the agency seeking it is precisely what the agency wants, standards and specifications have been prepared which set forth detailed requirements. The specifications which the product is to meet are set forth in the purchasing document. Failure to comply with the specifications is cause for rejection of the seller's proposal or quotation.

The Department of Defense (DOD) issues standards in three categories: Air Force–Navy Aeronautical (AN), Air Force–Navy Design (AND), and Military Standards (MS). Items which are not covered by the AN, AND, or MS documents are under the General Services Administration (GSA) procurement regulations and use GSA specifications. As one would expect, there are literally thousands of such specifications published and in effect.

In addition, many government agencies have regulatory functions. Examples are the Occupational Safety and Health Administration (OSHA), the Consumer Product Safety Commission (CPSC), and the National Highway Transportation Safety Board. The acts of Congress which established these agencies gave them the power to write and issue regulations to enable them to carry out their assigned missions. These regulations often cite existing published standards in their text. When that happens, the published standards are given the force of law. What may have been written as a voluntary standard is transformed to a mandatory standard, and the designer must be aware of this situation. Failure to comply with the standard can result in legal repercussions.

Regulations are initially published in the *Federal Register*. This publication is issued every day that the federal government is open for official business, and it runs to many thousands of printed pages each year. The first publication is in the nature of an official notice of what the publishing agency proposes to do in the way of regulations. The public at large is invited to comment on the proposed regulations and to attend any hearings which may be held regarding them. In this way a certain amount of public input is obtained. To that extent, such regulations are based on public consensus. After the comment period is over and all the hearings have been held, a final version of the regulations is published in the *Federal Register* with a date upon which the regulations are to become effective. To keep abreast of what is going on in the regulatory arena, it is highly recommended that designers make regular use of the *Federal Register*.

Once the regulations have been approved and published, they are collected into the *Code of Federal Regulations* (CFR). There are 50 CFR titles covering all aspects of the federal government. The CPSC is under Title 16, Commercial Practices (Parts 1000–1799); the National Highway Traffic Safety Administration is under Chapter III of Title 23, Highways; OSHA is under Title 29, Labor (Parts 1900–1999); Mine Safety and Health is under Title 30, Mineral Resources (Parts 1–199) even though this agency is part of the Department of Labor; the Defense Department is under Title 32, National Defense, but the Army Corps of Engineers is under Title 36, Parks, Forests, and Public Property (Parts 300–399); the Maritime Administration and parts of the Coast Guard are under Title 46, Shipping, while other parts of the Coast Guard are under Title 49, Transportation (Parts 400–499). Because of the complexity of the CFR, an *Index and Finding Aids* is also published.

Significantly, all the published regulations are reviewed and revised every year. It is advisable for designers to keep obsolete copies on file for reference in case of disputes which may arise at a later date. Having the correct wording of the applicable regulations as of a certain date can be very important in the event of litigation.

The formats for the specification documents issued by the various government agencies are all similar, differing only in the details. The specifications issued by military organizations are each identified by the capital letters MIL, followed by a single capital letter which is the first letter of the name of the material being specified. For example, the specification for Hydraulic Fluid has a MIL-H designation, followed by a number such as MIL-H-5606. If there have been any revisions to this specification, a letter is added after the number. The effective date of this revision is given on a line under the identifier. Below that are the identifier and effective date of the previous specification, to provide a paper trail in case of disputes. To continue the example, the complete identifier for the hydraulic fluid is

MIL-H-5606E
29 August 1980
Superseding
MIL-H-5606D
26 January 1978

The heading indicates that this is a military specification, and the name of the item is given with the noun first, followed by any descriptive adjectives. For the 5606 material the name is given as

Hydraulic fluid, petroleum base;
Aircraft, missile, and ordnance

This is followed by a statement of which departments and agencies of the Defense Department have approved this specification.

The body of the specification is set forth in six sections. The first section defines the scope of the specification in detail. The second section lists other documents which are referred to in the text and which become a part of the specification by reference. This frequently includes federal specifications, other military specifications, federal and military standards (which are different from specifications), and any applicable industrial or commercial standards and/or specifications. This list can be something of a challenge, since the referenced government specifications and standards will also include lists of more applicable documents. This situation is known as the *specification tree,* and it can grow very quickly into a totally unmanageable condition with literally hundreds of branches. Designers are strongly advised to find out how many levels of documentation references must be included. Any requirement for more than three levels of references should be questioned.

Section 3 of the specification finally gets down to the meat of the text under the heading *Requirements*. It is in this portion that the details are set forth which are intended to ensure that the agency gets precisely what it wants, and nothing else. The attitude of some vendors that "what you see is what you get" is completely out of bounds when the requirements section is involved. To make doubly certain that there are no deviations from the stated requirements, many materials are required to be tested and certified as meeting all the requirements. If they pass the qualification testing, the product is then entered on the *Qualified Products List* (QPL). There are many materials which are procured only from the QPL, and the designer must be aware of that fact to prevent accidental use of a nonqualified material.

Section 4 contains the *quality assurance provisions*. It usually includes a statement regarding who is responsible for the carrying out of the quality assurance (QA) program, either the government or the contractor. There are detailed requirements for the sampling plans to be used, and sometimes there is a requirement that the qualification testing be repeated at periodic intervals to ensure that the product is still in accordance with all the requirements. Such provisions can have a substantial impact on the costs of a proposed design, hence designers should read them carefully to avoid a possible unpleasant surprise later on.

In addition to the detailed requirements for the product itself, military specifications call out the details of how the product is to be packaged and how it is to be packed into that package. Section 5 gives the specified information, along with any special considerations such as preservation means, markings, and any necessary warnings. Section 6 contains miscellaneous information under the heading *Notes*. Of particular interest to designers is the part on ordering information, since it gives a clear-cut indication of what the procurement agency will be including in its requests for quotations. This information enables designers to make certain that all items of cost are included in their calculations. The notes often give background information regarding the intended use for the specified product, how it will be stored prior to use, and any special conditions that may be applicable.

Federal specifications follow the same format as the military specifications, with the same six sections and headings. The identifiers at the top right-hand corner of the first page have a slightly different style, using two or three repeated capital letters in place of the MIL designation, but otherwise it is the same. As an example, the specification for wire rope and strand has the heading

RR-W-410D
April 25, 1984
Superseding
Fed. Spec. RR-W-410C
September 18, 1968

Industry Standards

While the federal government's specifications and standards have a common format, industry standards are noted for their individuality. In the 1990s there are nearly 350 standards-generating organizations in the United States alone, each with the goal of writing standards for its area of interest and following its own approach to the task.

For many years designers had no central facility from which to obtain information regarding industry standards, and it was easy to overlook available documentation. The establishment of databases in recent years has resulted in the publication of the *Index and Directory of U.S. Industry Standards*. This two-volume set contains listings of more than 35,000 documents from more than 340 publishing organizations. Volume 1 contains the subject index and lists all applicable standards from all sources for any selected subject. It is an excellent first source for designers seeking to find out whether there are any pertinent standards already in existence and which organizations issue them.

Volume 2 contains a listing of all standards-publishing organizations in alphabetical order by their acronyms. The first listing is AA (Aluminum Association), and it runs through the WWPA (Woven Wire Products Association). Each organization entry lists all publications in numerical order. A society directory lists the name, address, and telephone number of each organization for ordering purposes. Significantly, most standards publishers have someone on staff who is well versed in the technical aspects of their publications. This person can often be helpful in explaining the history and development of the standards and the reasons for certain provisions and can give advice about the likely direction of future revisions.

As mentioned previously, periodic revisions of standards are routinely made to keep them in accordance with the latest advances in technology. Designers, who have to deal with the development lag time of their proposed products, must therefore obtain the most current information possible. The technical people at the standards publishers are highly recommended as the source for this information.

The index itself is published by Information Handling Services of Englewood, Colorado, and is available from Global Engineering Documents. Global has offices at 2805 McGaw Avenue, Irvine, CA 92714, and at 4351 Garden City Drive, Landover, MD 20785. Global can also supply copies of any desired document for a fee. With such a comprehensive database readily available, there is no reason and little excuse for any designer to be unaware of existing applicable standards for reference.

While most industry standards are of the consensus type, they are the result of input from members of an industry trade association only. As a result, there tends to be a certain bias inherent in the wording of such standards. Designers should be aware of this possibility and evaluate it accordingly. Even some of the larger standards-writing organizations are not immune from this problem. A major example involved the American Society of Mechanical Engineers (ASME), which prepares the well-known and widely accepted *Boiler and Pressure Vessel Code*. The ASME was successfully sued by the maker of a device used to indicate the liquid level in boilers. The basis for the suit was that the ASME code unfairly discriminated against his "Hydrolevel" device.

Some organizations, while not specifically directed to one type of product or service, have a particular focus to their interest. An example is the *National Safety Council*. This organization was founded in 1913, and it publishes books and pamphlets setting forth safety considerations for a wide variety of products. These publications are not standards but include many recommended practices. The NSC keeps statistics on accidents, including causation, frequency, and severity measures. One of the major functions of the NSC is to provide input to the organizations which do write standards, and they are heavily represented on ANSI and ASTM committees.

Another large and well-recognized nonindustry standards-writing organization is the *National Fire Protection Association* (NFPA). For many years the NFPA published documents known as *underwriters' standards*. These were used by fire insurance companies to reduce the degree of fire hazard inherent in many products or processes. Among the better known was standard 13 for the design and installation of automatic sprinkler systems. Standard 70 related to the installation of electrical equipment and is now known as the *National Electrical Code*. Many of the NFPA codes and standards have been adopted as national standards by ANSI. They are collectively published by the NFPA as the *National Fire Codes*. Annual revisions are provided for most standards.

Underwriters Laboratories is an independent, not-for-profit organization which was founded in 1894. It was about that time that commercialization of electrically generated power was becoming practical, and the spreading use of this form of energy revealed myriad new and unexpected hazards to life and property. UL was sponsored by the insurance underwriters "to establish, maintain, and operate laboratories for the examination and testing of devices, systems and materials to determine their relation to hazards to life and property." In carrying out the testing and examination portions of their mission, UL staff members prepare a *standard for safety* for each category of device. Each standard carries an identifying number along with the category name to which the standard applies. As an example, UL 727 is entitled *Oil-Fired Central Furnaces*. It has been adopted by ANSI, which formerly listed it as Z96.1 but now calls it ANSI/UL 727.

The format for UL standards always begins with a statement of the scope of the category being considered. This is worded quite precisely, and it should be consulted by designers to make certain that the selected standard is the correct one for the particular design task and includes the proposed device within its scope. As noted for other standards writers, UL reviews and revises its standards at periodic intervals, hence it is important that the applicable edition be obtained, including any revision sheets or additional pages issued subsequently.

While some standards writers use the specification style, laying down fixed and rigid requirements, and some others use performance criteria, leaving it to the designers to meet the requirement in any way they can, UL uses a mixture of both specification and performance. Details of construction, the type of electrical components used, and the spacings required for circuits carrying specific voltages are closely defined. The performance of the device is tested under a variety of foreseeable conditions in the UL laboratories before the device is approved. In addition, certain online tests may be required at the facility where the device is produced to ensure that the product remains in compliance with the requirements. The application of the UL label, which is protected by a trademark, is very carefully monitored to prevent possible abuse and loss of confidence in the value of that certification.

Industry standards relating to materials and their properties have been established and published by several organizations. Many ASTM specifications have been developed for the common metals and many alloys, giving values for physical properties and chemical composition. These standards are so well accepted that many purchasing documents call out the particular ASTM standard which must be met by the material.

Product Standards

Wood, which is one of the commonest building materials, poses special problems to the specifier since it is a product of nature and thus is subject to comparatively wide variation in physical properties. In addition, many different species of wood are suitable for some applications but not others. Selection of the proper material by designers must be made carefully, but fortunately several published reference standards offer excellent guidance. American Lumber Standards are issued by several regional wood products associations and contain their rules for the grading of particular varieties of woods. As an example, the Western Wood Products Association rules apply to Douglas and hemlock-true firs; Idaho white, lodgepole, ponderosa, and sugar pine; incense and western red cedar; western larch and Engleman spruce. The grading rules are established pursuant to *Product Standard PS 20,* which is a voluntary standard published under the procedures of the Department of Commerce for this product. The grading rules establish uniform lumber sizes, grade names, and grading provisions. This standard is called the *American Softwood Lumber Standard.*

Actual grading of individual pieces of wood is done by machine or by inspectors specially trained in this work and provided to the lumber-producing mills by regional lumber inspection bureaus. It is recognized that lumber grading cannot be considered an exact science because it depends to a certain extent on a visual inspection and a judgment call by the grader. However, years of experience and the consensus process have resulted in rules which are generally believed to result in a maximum of 5 percent below standard as the variation between individual graders.

Voluntary product standards are developed under procedures published by the Department of Commerce in Part 10 of Title 15 of the *Code of Federal Regulations.* The Voluntary Products Standards program is administered by the *National Institute of Standards and Technology* (NIST), formerly the National Bureau of Standards. NIST acts as the coordinator during the development of a standard, giving editorial and technical assistance and smoothing out disagreements among interested parties. NIST publishes the resultant standard as a public document. The functions and procedures are similar to those used by

ANSI and involve a large measure of consensus input. As with other standards organizations, reviews and revisions of the product standards are carried out when necessary.

Note that the purpose of the Voluntary Standards Program is to establish nationally recognized requirements for various products. Their use, however, is strictly voluntary, as the title says. NIST has no regulatory power and cannot enforce any of the provisions in the published standards. They do not have the force of law, by themselves, except when cited as a reference in legal codes, documents, or contracts.

The format for product standards starts with a statement of the purpose of the standard, followed by the scope of what the standard covers, a listing of definitions of terms used in the text, and then the technical requirements. Inspection and testing procedures to ensure conformance to the given requirements are given in section 5. This may be followed by identification call-outs and the members of the standing committee which acts to update the standard when needed.

Commercial standards (denoted by the initials CS in contrast to PS used for product standards) are also published by the Department of Commerce for items considered to be commodities. These are consensus standards which establish test methods, ratings, certifications, and labeling for manufactured commodities in order to provide uniform bases for fair competition. Like the product standards, commercial standards do not have the force of law by themselves, since they are purely voluntary. When incorporated in codes or contracts, they do acquire legal status. The formats for CS documents are very similar to those for the PS documents, and designers should become familiar with both types.

Building Codes

In 1905 the property and casualty segments of the insurance industry, through the National Board of Fire Underwriters, began the publication of a national building code. Their purpose was to improve life safety and fire protection features of buildings. The intent was to reduce loss of life and fire damage to property through prudent design considerations. Underwriters Laboratories had been in existence for 10 years and had developed the capability for testing and rating various types of building construction for resistance to fire. This permitted the specification of certain values for fire resistance in terms of how long the given construction would withstand the effects of attack by a standard fire. By specifying a suitable time duration for fire resistance of building components, the occupants of the building would be ensured an adequate amount of time to get out of the fire area. Of course, this also entailed the designing and specification of suitable means of exit.

The *National Building Code* was the one recommended by the American Insurance Association until it was combined with the code published by *Building Officials and Code Administrators International Inc.* (BOCA). This organization began in 1915 and publishes the *BOCA National Building Code.* It classifies buildings by both type of construction and by occupancy, and it uses a mixture of both performance and specification styles to establish requirements. When the BOCA code is called out in legal documents, the requirements have the force of law. Changes in requirements are incorporated in the code every 3 years through the periodic revision program.

The *Uniform Building Code* (UBC) has been issued by the International Conference of Building Officials since 1927. This code is very similar in its provisions to the BOCA code, and it follows the 3-year revision program. The conference also issues several other codes in addition to the one pertaining to buildings. This code is most widely used in the western portion of the United States and is administered through the conference office in Whittier, California.

The southeastern portion of the United States uses the *Standard Building Code,* which is published by the Southern Building Code Congress International, Inc. The office is located in Birmingham, Alabama. This code was first issued in 1945 and is the only one which makes revisions annually. They also publish several other standard codes which cover such topics as fire prevention, gas, amusement devices, swimming pools, excavation and grading, and plumbing.

All these regional codes are similar in their content and requirements. Obviously, designers must determine which code, if any, applies to their particular problems since there are slight differences among them.

In addition to the major regional codes, many large cities and some states have prepared their own building codes to reflect local conditions. In addition to the construction features of buildings, other codes may be in effect for plumbing, electrical, gas, and mechanical features. The local building department should always be consulted by designers to ascertain whether any special or unique requirements have to be met or whether approvals or certifications are required before certain materials or devices can be used.

Fire Codes

In 1896, about nine years before the first national building code was prepared by the American Insurance Association, the NFPA recognized the significant reduction of fire losses which could be realized by the installation of automatic sprinkler systems. To make certain that these relatively recent developments in fire protection were properly designed, installed, and maintained, NFPA prepared and issued a standard, setting forth detailed requirements for such equipment. It is still one of the most important NFPA standards, known as standard 13, and has been revised many times since its original version. In 1897, a standard for electrical installations was issued, reflecting the insurance underwriters' concerns with the shock and fire hazards connected with the rapidly growing use of central station generating stations and their distribution systems. In 1911 this standard, now known as NFPA 70, was formally adopted as the *National Electrical Code,* and it is included in the ANSI standards listings. Any designers whose products involve the generation or use of electric energy should be familiar with the detailed requirements set forth in the *National Electrical Code.*

Since its founding in 1896, the NFPA has prepared and issued hundreds of standards in addition to the sprinkler code and the electrical code. These are collected into a set of paper-bound volumes known as the *national fire codes.* They are revised annually. In addition, a set of loose-leaf binders is available through a subscription service which permits the assembled codes to be kept up to date by removing and replacing obsolete pages. The physical volume of information provided in the fire codes can be appreciated from the fact that codes and standards require 12 bound volumes and 8 loose-leaf binders. In addition, three loose-leaf binders are required for the recommended practices, manuals, and guides which represent good engineering practice. The index for all this material runs to more than 120 loose-leaf pages.

The format for NFPA documents begins with a section for general information, followed by a statement of the scope of the standard, its purpose, and a set of definitions used in the text. The detailed requirements given in the body of the text are a combination of both specification and performance type of call-outs. As an aid to designers and others, the NFPA has prepared a set of handbooks which give the standard requirements plus a set of explanatory notes next to each section. This format tells the user what the requirement is and the rationale behind it as additional background information. This format is particularly valuable for the more arcane sections of the *National Electrical Code.*

Designers should be specially aware of NFPA 101, which is the *Code for Safety to Life from Fire,* also known as the *Life Safety Code.* This standard was first issued in 1913 and related primarily to exit requirements for buildings and structures. In the early 1940s several fires resulted in severe loss of life, including a nightclub in Boston and hotels in

Atlanta and Chicago. As a result of the lessons learned from these disasters, the life safety standard was updated and formally issued as a code in 1948. Where it has been adopted, it has the force of law.

Like the ASTM and ANSI documents, NFPA codes and standards are the result of input from many sources. The organization has more than 32,000 individuals and more than 150 national and regional societies on its membership roster. Input from all of them is encouraged in order to form a realistic consensus. Persons who served on the committee which prepared the standard are listed in the front portion of the document, along with the organization they represent.

Sources for Codes, Standards, and Specifications

The *Index of Federal Specifications, Standards and Commercial Item Descriptions,* issued every April by the General Services Administration, Office of Federal Supply Services, is available from the Superintendent of Documents, U.S. Government Printing Office, Washington, DC 20402 (202-783-3238).

Specifications for items procured by the General Services Administration are available from GSA Specifications Unit (WFSIS), 7th and D Streets SW, Washington, DC 20407.

Department of Defense specifications and standards are available from the Naval Publications and Forms Center, 5801 Tabor Avenue, Philadelphia, PA 19120.

Documents issued by the National Institute of Standards and Technology (NIST) are available from both the Government Printing Office and the National Technical Information Service in Springfield, VA 22161. However, you must have the ordering number for the desired document. This is obtained from NIST Publication and Program Inquiries, E128 Administration Building, NIST, Gaithersburg, MD 20899 (301-975-3058).

The UL publications are obtained from Underwriters Laboratories, Publications Stock, 333 Pfingsten Road, Northbrook, IL 60062 (312-272-2612).

Publications of the American Society for Testing and Materials can be obtained from ASTM, 1916 Race Street, Philadelphia, PA 19103 (215-299-5585).

The American National Standards Institute is located at 11 West 42nd Street, New York, NY 10018. Publications are available through that office or by calling 212-642-4990.

The National Fire Protection Association is located at 1 Batterymarch

Park, Quincy, MA 02269-9101. Publications can be obtained from that office or by calling 617-770-3500.

Addresses and other information regarding documents issued by the more than 340 standards-generating organizations in the United States can be obtained from Information Handling Services in Englewood, Colorado, or through Global Engineering Documents located at 2805 McGaw Avenue, Irvine, CA 92714 and 4351 Garden City Drive, Landover, MD 20785.

Summary

The continuing development of the engineering professions has generated a large body of information which has been collected into accepted ways of solving design problems. These good practices are reflected in codes, standards, and specifications which are developed and published by hundreds of organizations. The information is readily available and should be utilized in the initial stages of the design process. While some codes and standards are advisory, many have been enacted into law. In such cases, designers have no choice of whether to follow the stated requirements.

8
Product Liability

For many years the marketplace was governed by the principle "Let the buyer beware." This was also known by the Latin phrase *caveat emptor,* which is translated in slang to "What you see is what you get." This situation implied that buyers had the obligation to inform themselves about the condition of the article being purchased, including any hazards that might come with it. Buyers got the benefits of the purchased article but also assumed the risks, if any, of the hazards which were included. If buyers failed to inform themselves well enough before making the decision to purchase the article, they might suffer an injury or incur damage to property as a result of an undiscovered or hidden characteristic of the purchase. That was just too bad.

Buyers should have been more aware, or more wary, during the buying process. If any action was taken to recover for the injury or property damage, it could only be taken against the seller of the merchandise. If the seller were a small business and had limited resources, any recovery would also be limited. The two parties to the buying/selling transaction were the only ones involved, hence the legal obligation between them was a private matter. They had what is known as *privity of contract.* Liability for injury or damage caused by the product was limited by the terms of that contract.

If the seller also happened to be the maker of the article, the establishment of who was at fault in the matter was relatively simple, since no other parties were involved in the handling of the product. However, as individual trades and crafts people were replaced by factory production of many items, an additional step was introduced in the marketing process. This was the use of organizations which purchased large quantities of product from the factory and then resold them to local merchants in smaller quantities. The marketing chain then changed from a maker-to-buyer system to a manufacturer-to-wholesaler-to-retailer-to-buyer system. Wholesalers and retailers did nothing to the product other than change its physical location. Whatever attributes were incorporated into the product by the manufacturer remained with it until it reached the hands of the ultimate purchaser.

The wholesaler had privity of contract with the manufacturer, the retailer had privity of contract with the wholesaler, and the buyer had privity of contract with the retailer. But the buyer had no privity of contract with the wholesaler or with the manufacturer, since the buyer did not have any contractual arrangement with either one. If a problem arose with the product, the buyer had recourse only against the retailer, just as under the old system. As before, if the retailer had limited assets, the buyer's chances of recovery were similarly limited. The buyer could not take legal action against either the manufacturer or the wholesaler, both of whom would have greater assets and thus a "deeper pocket" from which to recover.

The manufacturer and wholesaler were insulated from the adverse effects of product liability actions by the privity-of-contract concept. Even if the manufacturer was negligent in making the product and failed to exercise reasonable care to ensure that the product was not defective when it left the facility, the ultimate purchaser had no recourse against the manufacturer.

This situation came to a halt in 1916. A man named MacPherson bought a motorcar from his local dealer. The wheels on that car had wooden spokes, as was common at that time. After a comparatively short period of use, one wheel failed in service. Mr. MacPherson suffered a personal injury as a result of the wheel failure. It was determined that the cause of the failure was a defect in the spokes. The wheel had been defectively manufactured, and that defect had caused the wheel to fail prematurely. The judge found that the manufacturer had been negligent, and was held liable for the resultant injury. The reasoning was that "if the nature of a thing is such that it is reasonably certain to place life and limb in peril when negligently made, it is then a thing of danger." Since the manufacturer, not the dealer, had caused the danger to exist, the manufacturer was held liable for the injury regardless of the lack of a contract between the manufacturer and Mr. MacPherson. The privity shield was thus removed as a defense where negligence was the primary factor. From then on an injured party could take legal action against the negligent party regardless of that party's position in the chain of distribution.

With privity of contract eliminated as a defense in negligence cases, the issue of privity of contract in warranty matters was the next one to be attacked. Warranties come in two types. *Express warranties* are spelled out in detail in a contract. *Implied warranties* are not spelled out in detail, but are included in the contract by implication. Since both types of warranties are parts of the contractual agreement between the parties, it was long held that privity did apply when it was claimed that a breach of warranty was responsible for a personal injury or damage to property.

In 1961 another automobile case brought the end to the privity shield in warranty cases. One Ms. Henningsen purchased a new car from her dealer, drove it for a few days, and put less than 500 miles on it. While driving it, she heard a loud noise. Suddenly the steering wheel spun in her hands, and she lost control of the car. The car crashed into a wall, and Ms. Henningsen was injured as a result of the impact. Suit was brought on the basis that the manufacturer of the car had implied that the car was a merchantable product, i.e., it was of suitable quality for introduction into the marketplace for use in its intended manner. As a person with little knowledge about the design and construction of automobiles, Ms. Henningsen was not in a position to examine the car in minute detail to determine whether it was, in fact, of merchantable quality. She had to rely on the manufacturer's implied warranty of fitness. When the steering gear failed in service, it became obvious that the car was not fit for its intended purpose. The implied warranty of merchantability was breached. The court found that the implied warranty went with the product and accompanied it through the distribution channels of the marketing system set up by the manufacturer. The manufacturer was thus liable for any injury or damage resulting from the breach of the implied warranty, in spite of the fact that there was no direct contract with Ms. Henningsen. Thus the privity shield was wiped out as a defense against breach of warranty.

One result of this decision was that a breach of warranty began to be interpreted as an instance of strict liability. If a manufacturer placed a defective item into the stream of commerce and that defective item caused injury or damage to some person or thing, the manufacturer was strictly liable for the consequences of the defect, regardless of any contractual limitations which might exist. The limitations from legal niceties of contract law such as disclaimers, warranty restrictions, and privity were swept away. The theory was that a manufacturer owed a duty to the ultimate purchaser of the product. That duty was to not cause wrongful harm to the purchaser by placing a dangerously defective product on the market.

Causing wrongful harm to another person or her or his property is a civil wrong, rather than a contract violation. Such acts are called *torts*. The sweeping away of the contractual limitations on product liability opened the way for such cases to be treated under the laws of torts. Under the laws of torts, the injured party is entitled to compensation from the party performing the wrongful act. Since civil wrongs have been going on for millennia, the law of torts is a well-developed field with a huge body of applicable cases.

The definitive shift from contract to tort law in product liability cases took place in 1963. A homeowner purchased a piece of powered woodworking equipment for his home workshop and used it without incident for more than a year. He then purchased some attachments for the equipment which enabled it to be used as a lathe. One day he was turning a piece of wood, using the lathe attachment in its normal manner, when the workpiece flew out of the machine and struck him in the head. He was injured seriously. Several months later he decided to sue both the manufacturer of the equipment and the retailer who sold it to him. The grounds for the suit were both negligence and breach of warranties.

In considering this case, the court came to the conclusion that contract law was no longer the proper theory of liability. The decision was made that the tort theory was the correct way to treat such matters. The decision was based on the belief that the purpose of product liability is to ensure that the costs of injuries resulting from defective products are borne not by the injured parties, but by the manufacturers of such products. The laws relating to sales, with all their intricate details based on contractual relationships, were held not to be the controlling factors. The facts in the case showed that the injury took place while the equipment was being used in a manner for which it was intended. Nothing that the plaintiff did contributed to the causation of the mishap. From this, it followed that the machine was not properly designed and manufactured to perform its intended functions in a safe manner. The plaintiff was unaware of the defects since they were not readily perceivable and since no warnings were given about the possible dangers involved. The court concluded that the plaintiff had been wronged by the manufacturer. The manufacturer had committed a wrongful act, a tort, and was held liable for the consequences.

Although it took from 1916 to 1963 to establish it, the theory of tort law in product liability matters is now firmly in place. The American Law Institute, an organization which codifies the results of case law decisions, issued a definitive summary of current opinions on product liability in 1965. This summary was included in the *Second Restatement of the Law of Torts*. It reflects the attitude of society as contained in the judicial decisions handed down in recent years.

Since tort law is a complex subject, the *Second Restatement* contains many sections and paragraphs. The one which is most pertinent to the product liability issue is section 402A. It states:

> 402a. Special Liability of Seller of Product for Physical Harm to User or Consumer
>
> (1) Who sells any product in a defective condition unreasonably dangerous to the user or consumer or to his property is subject to liability for physical harm thereby caused to the ultimate user or consumer, or to his property, if
>
> (a) the seller is engaged in the business of selling such a product and
>
> (b) it is expected to and does reach the user or consumer without substantial change in the condition in which it is sold.
>
> (2) The rule stated in Subsection (1) applies although
>
> (a) the seller has exercised all possible care in the preparation and sale of his product and
>
> (b) the user or consumer has not bought the product from or entered into any contractual relation with the seller.

Translated from legalese to plain language, section 402A simply says that manufacturers are *strictly liable* for any harm which their products cause to the person at the end of the marketing chain. The liability applies even though *all possible care* was used in preparing and distributing the product. If a clunker somehow gets through the quality assurance system, no matter how good that system may be, and a mishap occurs as a result, the manufacturer is going to pay the injured party for the damage. In many cases, the manufacturer will pass the risks of such liability claims to an insurer in return for payment of a suitable premium.

The term *preparation,* as applied to a product by section (2)(a), includes the effort used in designing the product as well as the actual making, inspecting, and packing of it. If there is a defect in the design of the product, that design defect can render the product unreasonably dangerous and result in the manufacturer's being held liable under section 402A. Although the original version of section 402A related specifically to ultimate users or consumers of a product, courts have subsequently held that its principles also apply to persons who are innocent bystanders. Even though they were not actually involved in the chain of distribution and never used or consumed the product, if they or their property suffered injury or damage, innocent bystanders are considered to have been wronged and can recover from the manufacturer. Similar extensions of strict liability in tort have been made to the suppliers of components which are incorporated into other products without modification by another manufacturer. People who lease products to others, rather than selling them, have also been put under the strict liability umbrella.

The switch from privity of contract to strict liability in tort has changed the marketing maxim from *caveat emptor* to *caveat venditor.*

Instead of the buyer having to be wary, now it is the seller's obligation to exercise caution. As mentioned previously, one of the cautionary measures taken by manufacturers is to get an insurance company to assume the risk of any loss and pay the company a premium for doing so. The coverage is called *product liability insurance*. The premiums paid to the insurance company are part of the manufacturer's costs of doing business. The manufacturer adds these costs to the rest of the costs of production as an expense item. The cost of product liability insurance is passed on to the wholesaler, the retailer, and finally the ultimate consumer as a price increase. Thus the little person at the end of the distribution chain winds up paying for the costs of the manufacturer's mistakes. Through the use of the insurance principle, the costs of the risk are spread over many people rather than being concentrated with only a few injured parties. However, the costs are still there.

Note that the costs of medical treatment for a personal injury or replacement of damaged property are only the direct product liability costs. Many indirect costs must also be considered. These include the underwriting and administrative costs of the insurer and the costs of any litigation which may ensue. If the litigation can be settled without going to court, there are attorneys' fees and expenses which must be paid. A substantial portion of the settlement goes to the injured party's attorney if the case was taken on a contingency basis. If the matter cannot be settled by negotiation but goes to court, there are additional costs for court time, the costs of taking statements or depositions from various witnesses prior to the trial, and the fees of any expert witnesses whose testimony is used in court. Generally speaking, litigation is a very expensive business that adds considerably to the overall costs of product-caused injury or damage.

In some instances, the insurers have experienced so many liability claims against certain products that they have had to raise the premiums to a very high level. Then the producer of that product is forced to consider the costs of product liability not as a minor item, but as a major component of the overall cost of the product. Product liability then commands attention from management as an expense item which can be controlled. If certain corrective actions can be taken and the frequency and severity of the claims are reduced, the insurance premiums may be brought down to an acceptable level again. And who takes the corrective actions? Often it turns out to be the designers who overlooked a hazard which could have and should have been designed out of the product originally. In other cases the excessive claims may be the result of low quality which requires improvements by manufacturing. However, the key consideration in all such instances is that the excessive costs of product liability are largely avoidable.

Occasionally a product engenders such high product liability claims that the insurance companies refuse to take on the risk at all. They simply refuse to underwrite the liability, and the product thus becomes uninsurable. When that happens, the manufacturers have two choices. They can withdraw the product from the market, or they can continue it in their product line but without insurance protection. The second alternative is called *going naked*. Any product liability losses must be paid by the manufacturers from their own assets. While large companies may be able to afford to carry their own insurance, smaller ones could be forced out of business by one large adverse judgment.

In addition to the previously mentioned direct and indirect costs of product liability, substantial amounts of money may be involved if the jury determines that the injury or damage was caused by willful actions on the part of the manufacturer. The amounts awarded are called *punitive damages*. They are commonly several times as large as the ordinary or compensatory damages, and they are intended to punish the wrongdoer.

Product Liability Insurance and Rates

The risks of loss from product-caused injuries or damage are assumed by insurance companies which engage in the general liability field. At present there are several thousand such companies in the United States alone and more in other countries. Thus competition is great.

General liability coverages include many things other than products. Products are placed in the commercial general liability category and cover manufactured products and completed operations. A very large amount of data regarding the frequency and severity of commercial general liability losses has been collected by the insurance companies over many years. These data are reported to the *Insurance Services Office* (ISO), which is headquartered in New York City. The ISO maintains the database for more than 1500 different classes of products and types of operation. The experience reflected by the data is used by the ISO to establish a recommended rate for each of the 1500 classes. These rates are intended to produce enough premium income for the insurance companies to let them pay the expected claims, the expenses of settling claims, and their general and administrative expenses.

The ISO rates are published in a thick manual which is available to insurance agents and underwriters at various companies. In addition, they are filed with the insurance regulatory bodies of the various states. The rates encompass two components. The large part is for bodily injuries to persons, called *BI coverage*. The second, and much smaller part, is for damage to property, called *PD coverage*. The sum of the two parts gives the total rate.

The published rate for manufacturing enterprises is given in cents and fractions of a cent for each $1000 of gross sales of the product by the manufacturer at the factory price. For wholesale distributors and retail merchants, the rates are also based on sales at their prices. The range of rates is quite wide. One of the lowest is for ISO class 50493, which includes fruit, vegetable, and grocery wholesalers. The BI rate is 0.049 cent per $1000 of gross sales, while the PD rate is 0.003 cent per $1000. The highest rate is for plants which refill carbonated water siphons. The rates are based on each 10,000 fillings rather than gross sales, but the BI rate is more than $60 and the PD rate is more than $3 for the product coverage.

In some instances, the ISO database is not large enough to permit a realistic rate to be calculated. This is noted by a parenthetical entry of the letter (a), which tells the user of the rate manual to refer to the home office for information. Actuaries and underwriters at the home office may have some data applicable to the particular product being rated, and they formulate what is known as a *guide (a) rate*. If very few data are available, the underwriter will establish a *pure (a) rate*. This rate is based on the underwriter's knowledge and experience of the rates for similar products and what other companies are charging. Since there are so many liability insurance companies in business, the underwriter must be careful that the rate she or he arrives at is realistic and competitive.

In some instances, however, the underwriter may conclude that the product or the company which produces it involves such high risks that the business will not be acceptable to the insurance company. The result may be a decision not to offer coverage at all or to set such a high rate that the manufacturer will turn to another insurer for coverage. In his or her position, the underwriter clearly has major authority and responsibility for establishing the rates which the company will charge.

In making decisions, underwriters at the larger insurance companies rely on information obtained for them by members of their loss prevention staff. These representatives visit the manufacturer's plant, sometimes several times per year. During those visits they examine the quality assurance programs being used, evaluate the hazard recognition efforts of designers, check on the manuals and instructions which accompany the product into the marketplace, establish the history of the product, get data on the frequency and severity of any injuries caused by the product, and evaluate the overall ability of management. A report is prepared by the loss prevention people and given to the underwriters to assist them in their assessment of risk.

Although many loss prevention staff members have engineering training, they do *not* make recommendations to the engineering department of the manufacturer prior to the production and marketing of the product. Obviously, if the insurance company engineer made a recommendation and the manufacturer implemented it, any liability claims on that product would be tainted by the engineer's input. To avoid this conflict of interest, the insurance companies simply do not do "upfront" engineering.

Once the product has been introduced into the marketplace and some data have been accumulated on the frequency and severity of claims, then and only then will the insurance companies get into the picture. Based on the collected experience, recommendations for changes are formulated and submitted to the manufacturer. The intent of these recommendations is to prevent future losses by eliminating their causes. For that reason, such technical professionals are called *loss prevention engineers*.

Implementation of the loss prevention engineers' recommendations is still at the discretion of the manufacturer. If the manufacturer does not choose to make the recommended changes, the underwriter may reflect this fact in the rate.

Product Liability Litigation

As pointed out previously, manufacturers of products can shift their liability for product-caused losses to an insurance carrier. In return for payment of a premium to the carrier, the insurance company assumes the risks of loss. On the other side of the equation, the user or consumer of a product has only limited opportunities for shifting the risk. Health and accident insurance can be obtained to cover personal injuries, and loss or damage to property can also be insured. Of course, both types of insurance require the payment of a premium for the coverage. Depending on the wording of the policy, the insured person will be compensated for the cost of any direct losses. However, indirect losses such as the ability to perform certain physical functions may not be covered under the personal accident insurance, and the property damages are often limited to the depreciated value of the article involved. Even with insurance protection, the person on the end of the distribution chain can still suffer uncompensated losses from damage or injury occasioned by defective products. Obviously, those who have no insurance protection are exposed to even greater risk of loss.

If an uncompensated loss to a person can be attributed to a tortious act on the part of the maker of the product or service which caused the loss, then a legal action can be brought by the injured party. Litigation attorneys handle such matters, often on a contingency basis. They take the case without any initial payment of a retainer fee. Their compensation is contingent upon their winning an award for the client and is usually an agreed-upon percentage of the award. This arrangement makes it possible for a poor person to hire an attorney, even though that person has no money for the usual legal fees.

Attorneys who handle such cases are usually members of the *Association of Trial Lawyers of America* (ATLA). The offices of this organization are located at 1050 31st Street NW, Washington, DC 20007. ATLA was founded in 1946, shortly after the end of World War II, and was originally known as the National Association of Claimants Compensation Attorneys. It now has about 70,000 members with 51 state and 5 local groups. Advancing the causes of persons seeking redress for damages against person or property is among their several objectives. In addition to maintaining a sizable library, ATLA conducts research on insurance, product liability, and several other areas of legal interest. ATLA has more than a dozen sections, including ones on commercial litigation and tort law.

As the largest of the advocacy groups, ATLA is actively engaged in publishing pertinent materials. Among these are the *Product Liability Law Reporter* and the *Professional Negligence Law Reporter,* both issued 10 times per year. In addition to the periodicals, they publish books, audio cassettes, videotapes, and other media products devoted to civil and criminal trial advocacy. With a staff of about 150 persons, this organization is obviously very well established and is extremely active in providing assistance to attorneys who represent plaintiffs in product liability matters. By any measure, they are a formidable opponent to any producer of a product which contains a recognizable defect.

Also note that several other smaller organizations, in addition to ATLA, are devoted to the field of trial advocacy. These include the American Board of Trial Advocates, which seeks to preserve the jury system; the National Board of Trial Advocacy, which certifies the competence of civil and criminal trial advocates; and the National Institute for Trial Advocacy, which trains lawyers for skills in trial advocacy.

The insurance companies, of course, regard the trial lawyers as enemies. When a product liability suit is filed against one of their insureds, the provisions of the policy obligate the insurance company to defend the policyholder against such actions. Considerable legal time and expense may be involved. Some part of the premium paid by the insured must be spent to cover the costs of these defensive efforts, but a good defense will serve to reduce the amount of compensatory damages awarded to the claimant and may eliminate them altogether if the insured party is found not to be liable for any of the alleged injury.

It is common practice for the plaintiff's attorneys to sue everyone in the distribution chain for the product involved. They do this to make certain that they reach the party actually responsible for the claimed wrongful act. If there is the slightest possibility that a given party can be brought into the suit, the party is added to the list of defendants, just to make sure that no possible responsible party is overlooked. Of course, often the named defendants must engage legal counsel just to prove that they were not liable for the alleged damages. Their insurers, if they have one, provide for their defense, absorbing the costs of the necessary legal filings, the hiring of experts, taking of testimony, and all other requisites to a proper defense. When the insurers are successful, the insurer is absolved from liability and let out of the case by the plaintiff's counsel. Since most defendants were originally firmly convinced of their total innocence, it is easy to see why they and their insurers regard such suits as frivolous.

An example of this situation arose in a case in which the author was called as an expert for one of the defendants. The incident involved a personal injury to the driver of a forklift truck. He was loading some merchandise onto a delivery truck, which was backed up against the loading dock at a warehouse door. The height of the truck bed was several inches higher than the height of the floor in the warehouse. However, the dock was fitted with an adjustable ramp which was hinged at its rear edge so that it could be tilted upward to provide a smooth incline from the warehouse floor onto the truck. The forklift driver drove to the ramp with a load on his forks, but had trouble getting up the slope because of the heavy load and the steepness of the incline. After a couple of unsuccessful attempts, he backed up a bit and took a flying run at the ramp. His front wheels hit the rear edge of the ramp so hard that the impact broke the hinge which attached the ramp to the building. The whole ramp, with the driver and his load, moved forward and fell into the pit under the ramp. The driver was injured in the fall.

The attorney for the driver sued the maker of the ramp, the people who installed it, the maintenance firm which had serviced it, and several firms which had supplied components to the ramp manufacturer. Among the component suppliers named as a defendant was one that made the electric motor which supplied power to raise and lower the ramp. To any competent engineer, it was immediately obvious that the motor had not contributed to the causation of this injury in any manner. The expert's report convinced the plaintiff's counsel of that fact, and the motor supplier was dropped from the case. Still, at the time of the onsite examination of the loading dock, the motor manufacturer had to pay for the services of the examining expert, the legal counsel who witnessed the examination, and two technicians from the factory. Considering the expense involved in defending this totally unwarranted lawsuit, it is no wonder that such legal actions generate antagonism between opposing sides. Regrettably, this waste of resources is a fundamental consequence of the adversarial nature of the U.S. legal system and is not likely to change. The costs are reflected in increased premiums charged for product liability insurance or out-of-pocket expenses to those who do not carry such insurance.

Consumer Organizations

Through the trial advocacy support provided by the ATLA, many injured parties have been assisted in their efforts to obtain compensation for losses or damages. In the eyes of some, the judgments assessed against the defendants have served not only to compensate the plaintiffs for their claimed loss, but also to draw the producer's attention to the fact that there was something about the product which needed improvement. In this way the plaintiff's lawyers feel that they are assisting in the improvement of product design. They have a point.

In addition to the legal professionals, who operate as practitioners, there are several formal organizations whose primary purpose is the promotion of causes favorable to consumer interests. One of the most prominent is a government agency, the Consumer Product Safety Commission. It is an independent agency, established by an act of Congress in 1972 to take over certain functions formerly handled by the Food and Drug Administration. A copy of the Consumer Product Safety Act is contained in Appendix A. Congress directed the commission "to protect the public against unreasonable risks of injuries and deaths associated with consumer products." This is a very broad mandate, indeed. The CPSC is not part of any other government department, does not have Cabinet status, goes to Congress with its own budget request, and reports to Congress rather than to the President. Because of its independent status and broad mandate, the CPSC has considerable power.

Among its greatest powers is that the CPSC can order a defective product recalled from the marketplace. For example, toys found to have sharp points which could injure small children or small parts which could be swallowed and choke such a person were determined to present an unreasonable hazard. Existing articles were ordered taken off the market, and further production was prohibited. The CPSC's power to ban a product is a severe threat to the inept designer or imprudent manufacturer. According to its data, the CPSC ordered more than 1500 recalls between 1973 and 1987. More than 220 million articles were rendered unsalable.

Another power of the CPSC is the preparation and promulgation of product safety standards. In the just-mentioned instance of the toys with sharp points, the question logically arises, How sharp is sharp? The CPSC staff produced a lengthy document to answer that question and developed specific criteria for the acceptance or rejection of suspected products. The standard became one of the CPSC *mandatory* standards. In some instances, power lawn mowers for one, the CPSC worked with the industry association in the development of a suitable *voluntary* safety standard. This was eventually published by ANSI under the sponsorship of the Outdoor Power Equipment Institute (OPEI).

There are some consumer products for which the CPSC has determined that no feasible standard would adequately protect the public against an unreasonable risk of injury. Several items, mostly of a chemical nature (some drain cleaners, loud fireworks, highly flammable water repellants) have been banned from sale in the United States. However, most mechanical and/or electrical products have had their unreasonable hazards taken care of through the voluntary standards method.

A key element in the CPSC operation is the gathering of information relating to the nature, frequency, and severity of injuries caused by consumer products. A huge database has been collected through the *National Electronic Injury Surveillance System* (NEISS), which gets its data from hospital emergency room admissions, death certificates, and medical examiners' reports. Analysis of the NEISS data enables the CPSC to identify existing hazards and to determine whether they should be classified as "unreasonable."

In-depth investigations of certain types of mishaps are conducted when the severity of the hazard justifies this additional effort. The results of these investigations are intended to permit the assessment of the degree of hazard and to suggest methods for reducing that hazard. Through this feedback procedure, the CPSC influences the design process for consumer products.

When a new consumer product is introduced to the market or a new use is found for an existing product, the CPSC gets into the act by identifying the hazards through the NEISS system. Obviously this effort is made only after the product has been designed, produced, and distributed to consumers. At present the CPSC has no method (other than the promulgation of product standards) for working with manufacturers during the design stages of new product development. Like the insurance companies, the CPSC does no up-front engineering or consultation. However, the CPSC's standards are readily available and should be consulted by any competent technical person during the initial design stages of any proposed consumer product.

Other materials, in addition to the standards, are available from the CPSC through its public information program. A set of fact sheets, intended for consumers, and many brochures and pamphlets are available upon request to any CPSC office. The CPSC also maintains a hot line which can be used to obtain information directly from the Washington, D.C., headquarters by calling (1-800-638-CPSC).

From the product liability viewpoint, it is important for product designers to know that this information is readily available to them and that it is available to members of the legal profession who may wish to use this material against them in court. Obviously, if the CPSC has determined that a given product presents an "unreasonable risk of injury to the public," it follows that that product contains a design defect. Successful defense of a product liability suit is practically impossible under those circumstances.

In addition to the CPSC several other federal agencies influence the design of consumer products. For example, the Food and Drug Administration (FDA), which is part of the Department of Health and Human Services, enforces radiation safety standards for such products as *x*-ray equipment, lasers, color TV sets, microwave ovens, and sunlamps.

Another example is the National Highway Traffic Safety Administration (NHTSA). This agency is part of the Department of Transportation and has jurisdiction over both consumer and commercial products which are used as vehicles. This includes passenger and commercial automobiles, trucks, buses, recreational vehicles, motorcycles, mopeds, and bicycles. In addition, it covers all the accessory equipment, plus car carriers and baby seats which are used in cars.

The NHTSA both writes and enforces performance-based safety standards. These standards establish minimum performance levels for both vehicle components and the vehicle as a whole. As is well known, NHTSA receives complaints from vehicle operators regarding alleged defects. It evaluates the complaints to determine what, if any, corrective actions may be necessary. It has the power to require a recall of vehicles if it is established that a defect exists.

In a manner similar to the CPSC, NHTSA also maintains a public information function. Public advisories and protection bulletins are published at regular intervals, and fact sheets are issued on current automotive problems. All this information is in the public domain. Therefore it is imperative that designers of products which are even vaguely related to the transportation field obtain all NHTSA standards and regulations applicable to their particular problems.

While NHTSA has jurisdiction over ground-based vehicles, safety standards for aircraft are established by the Federal Aviation Administration (FAA). This agency is also a part of the Department of Transportation. The FAA standards are mandatory for the construction of aircraft, the flammability of the interior finishes, and aircraft safety in general. The FAA has responsibility for certification of the airworthiness of various aircraft used for commercial or passenger service. Obviously, if the design does not conform to the established requirements, no certificate of airworthiness can be issued. In this way the FAA exerts a strong influence on aircraft design.

The designs for waterborne vehicles are regulated by the Office of Boating Safety, which is a part of the U.S. Coast Guard (USCG). The USCG, like the FAA, is a part of the Department of Transportation. The Office of Boating Safety sets safety standards which are mandatory for boats and associated equipment. Another USCG organization, the Office of Merchant Marine Safety, tests and approves items of marine safety equipment which consumers may purchase directly. This includes such items as life jackets (known to the USCG as personal flotation devices), fire extinguishers, and carburetor backfire flame arresters. Also, because of the possible effects of contamination and pollution on watercourses, the Office of Boating Safety tests and certifies marine sanitation devices.

The federal Department of Housing and Urban Development (HUD) also has a major interest in consumer affairs. Its primary interest lies in the quality of mobile homes. HUD has the responsibility for setting standards for the basic structure of mobile homes, and designers must comply with such regulations. Oddly, the safety requirements for the interior furnishings of mobile homes are established by the CPSC.

One additional federal agency, the U.S. Office of Consumer Affairs, has been set up to coordinate and advise various agencies on matters of interest to consumers. It represents the consumer viewpoint in federal agency proceedings and may contribute input to regulatory hearings. These hearings often influence the rules which are issued by the various agencies. In that way the Office of Consumer Affairs can have an important affect on the requirements which product designers must consider in their efforts.

Among the nongovernment organizations which have an active interest in consumer affairs, one of the most aggressive is the *Consumer Federation of America* (CFA). It is located at 1424 16th Street NW, Washington, D.C., and acts as an advocacy organization to promote consumer interests with Congress and various agencies of the federal government. For example, the CFA follows the injury information data published by NEISS. When a situation arises which the CFA considers to present an unreasonable risk of injury, the CFA files a petition with the CPSC, requesting action to address the described hazard. The CPSC considers the petition, and if the request is granted, a notice is published in the *Federal Register*. The notice states that the CPSC is beginning a rulemaking proceeding and that regulations may be issued by the CPSC as a result of this proceeding.

Publication of such a notice in the *Federal Register* should serve as a warning to all manufacturers of the product under evaluation. The notice includes an invitation to any interested parties to send certain specific types of information to the CPSC for consideration. It behooves every affected manufacturer to respond to this invitation, both for its own self-interest and to obtain information regarding the reasons for the proposed rulemaking. Note that publication in the *Federal Register* constitutes legal notification. However, no individual notices are sent out to manufacturers by the CPSC. The producers of products are expected to keep themselves informed. Clearly, reading the *Federal Register* on a regular basis is an important requirement for any organization which hopes to keep current in its industry.

In addition to filing petitions, the CFA works with the CPSC in the formulation of proposed rules. Rather than being confrontational, the CFA takes the approach that the development of voluntary standards is the preferred rule-making method. This way of dealing with the rule-making procedure brings the manufacturers of products into the process at an early stage. Manufacturers are thus able to introduce factors which they deem to be pertinent and have them considered by the CPSC. The voluntary standard which results from these efforts may be published by an organization which is entirely independent of the CPSC.

An example of this process is a voluntary standard which was developed by the CPSC and the toy industry, with assistance from the CFA. It was published by ASTM. It is ASTM Standard F963-86, entitled *Standard Consumer Safety Specification on Toy Safety*. The CFA has also worked on the voluntary standard for bunk beds, with assistance from ASTM committee F 15, and is currently (1991) working on a revision of the toy safety specification.

Another widely known and long-established consumer organization is *Consumers Union* (CU). It was originally a part of Consumers Research, but in 1936 it branched out on its own. The CU is located at 101 Truman Avenue, Yonkers, NY 10703, having moved from Mt. Vernon early in 1991.

Consumers Union has two primary functions: to test consumer products and to publish reports of the testing and formulating evaluations of the tested items. The CU compares and contrasts similar items, which their shoppers purchase in the open market, just as a consumer would do. The product evaluations are summarized in the form of recommendations. The CU publishes a listing of the tested items, stating which is best from the consumer's point of view.

CU also prepares and issues consumer advisory information, including topics on health, safety, and consumer services. The CU also publishes a syndicated news column and has a regular radio program as part of its public information efforts.

Sometimes CU uncovers a design defect during its product testing program. It is CU's practice to convey this information to the product manufacturer for possible corrective action. Also CU works with the CPSC on the testing and evaluation of certain classes of products and supplies product testing results to other government agencies or private organizations on request.

With a staff of about 350 persons and a large circulation base, CU carries a lot of weight in the consumer affairs field. It is militantly independent, and its publications accept no advertising. Because of its reputation, CU's published opinions command respect when cited in a product liability action. Product designers are well advised to respect any adverse comments they may receive as feedback from a CU testing program.

Summary

Since 1916 the field of product liability has changed dramatically. The old privity-of-contract defense against such liability has been abolished through court decisions and has been replaced by the doctrine of strict liability in tort. For protection against the risk of adverse product liability judgments, manufacturers have turned to insurance companies which assume the risk in return for payment of a premium. The premiums reflect the loss experience of the product. Many insurance companies provide technical advice to manufacturers as a means of improving product designs and thereby reducing product liability losses.

While manufacturers can shift the burden of losses to insurers, consumers must resort to the courts to obtain compensation. Such legal actions can be very costly and may involve considerable unnecessary expense. Several organizations espouse the causes of consumers and endeavor to improve the design of products through the writing of standards, dissemination of public information, and consumer advocacy. Product designers must be aware of these organizations and their efforts.

9

The Legal Process

One of the most feared four-word statements is *You have been sued.* It certainly commands attention. It indicates that you or someone in your employ has done something which somebody else feels is wrong, and that person has decided to take formal action against you to recover for that alleged wrong.

Complaints and Filings

The first stage in this process is the filing of a formal complaint with the local authorities. This document lists the parties to the complaint and sets forth the specific items of complaint. The wording of complaints is something which lawyers take a certain amount of pride in. They use phrases such as *willfully, wantonly, and maliciously* to describe an action which allegedly caused *grievous, serious, continuing, and permanent* injury to some person or thing. Such strings of words are part of the lawyer's stock in trade and are almost in the nature of what is commonly called *boilerplate,* since they are used over and over. While you need not be intimidated by the ferociously aggressive tone of the language, your best response is to engage the services of a competent attorney.

Discovery

Your legal counsel will file a reply to the complaint, admitting certain alleged facts but denying most of them. Your lawyer prepares a list of questions which are sent to the complainant's attorney, requesting specific facts to support the allegations made in the complaint. This is the first usual step in what is known as the *discovery process*. The list of questions is called an *interrogatory,* and it demands a wide variety of information. Lawyers often use prepared lists of questions, again a sort of boilerplate. The replies to the questions should tell your counsel the details of the complaint and permit a careful evaluation of the seriousness of the alleged wrong.

It is common for plaintiff's counsel to file a list of interrogatory questions to you, the defendant, also. You must reply to those questions, working with your counsel in the formulation of the replies. You may be astonished at the breadth of the questions and distressed at the amount of information you are being requested to provide. This can be particularly annoying when the requested information is many years old and has been put into storage at some remote location. If the information is of a proprietary nature or involves trade secrets, this issue can become particularly troublesome.

After both sides have submitted their lists of interrogatory questions and the replies have been received by both sides, most of the important details of the issue will have been disclosed to each of the adversaries. At that time an assessment of the parties' positions can be made and a decision reached whether to continue with the suit or to try to settle it without going to trial.

In many instances where there is a large amount of technical input to the matter at hand, outside consultants may be called in to provide the technical information required and present it in the form of a report. These consultants are commonly called *experts* because not only can they provide factual information, but also they can evaluate the relevance of that information and render opinions regarding its application to the issues at hand. Ordinary, nonexpert witnesses are not allowed to give any information other than the facts, as they know them. They cannot give opinions about those facts.

Reports prepared by experts and submitted to their clients may be "discoverable" items in the litigation process. If demanded by the opposing side, such reports must be produced for examination and evaluation. As part of the discovery process, the experts' reports provide important information to both sides. When added to the information obtained through the interrogatory process, the experts' reports complete the package of data with which the two opposing lawyers must work in attempting to reach a settlement. As expected, opposing experts often make very different evaluations of the facts given to them. The legal counsels must wrestle with that situation in trying to reach a settlement, and they may fail in their attempts.

Depositions and Examinations before Trial

If no settlement terms can be reached, further efforts at information gathering are begun. This involves the interviewing of various witnesses, either through depositions or examinations before trial. Both these procedures involve the taking of sworn testimony in front of a recorder, who takes down every word spoken. The testimony, being taken under oath, must be truthful; otherwise, the person giving the responses may be subject to the penalties for perjury.

The actual procedure for giving a deposition is quite straightforward. The witness is requested to appear at a certain place at a certain time. The place may be a lawyer's office, the place of business of the person giving the deposition, or any other mutually agreed-upon place. The attorneys for the opposing sides are present, along with the reporter.

The reporter usually uses a machine to record the spoken words, although there are still a few recorders who use a pencil and take shorthand on a notepad. It is important for the deponent (the person being questioned) to remember that the recorder can write only spoken words and cannot record gestures, smiles, or tones of emphasis in speaking. Remember that the transcribed record will consist of only the utterances at the deposition, there being no video component to reveal shakes or nods of the head, pointing of fingers at places on a drawing, or what a photograph may show. Particular attention must be paid to the grammatical matching of pronouns and their antecedents. Reading some deposition transcripts can be a maddening exercise in trying to match up a *that* with a particular thing or a *he* with a particular person. Video depositions are now accepted by some courts and may be shown to the jury at the time of trial.

Depositions or examinations before trial consist of a long series of questions asked by one of the lawyers, with answers provided by the deponent. Before the questioning begins, the lawyer explains the deposition process and includes an admonition that if the deponent answers the question, it is presumed that the question was understood. That is not always the case, particularly if the question is a long and convoluted one. If that happens, the witness should simply state that she or he does not understand the question and ask the questioner to rephrase it.

The first few questions are usually devoted to getting the deponent's name, address, and certain vital statistics into the record. After that, some information regarding life history, education, places of residence, and connection with the matter at hand is requested. After these baseline questions are answered, the specific questions begin.

Since the legal matter is a civil one involving a dispute between two adversarial parties, there are no detective story type of questions such as "Where were you on the night of January 13th?" The objective of the person taking the deposition is to find out what the deponent knows about the matter at hand. There is emphasis on the word *knows* to the exclusion of what the person *feels* or *thinks*.

Many questions are carefully crafted so that the answer can be given as a simple yes or no. If it is possible to answer such a question truthfully with a yes or a no, do just that, without adding further words of explanation. The general rule during deposition testimony is to *never volunteer any information*. Just answer the question and only the question.

Sometimes a simple yes or no is not the whole answer to the question. In that case the witness, unless directed to answer yes or no, should explain why the simple answer is not suitable. There are few purely black-and-white situations in litigation, so that when a gray area comes up, it is entirely proper to fill in the necessary details in giving an answer.

Compound questions, in which more than one thing is asked, are particularly dangerous. The questioner should be asked to rephrase the question. Similarly, an apparently clear question can have a completely ambiguous answer. As an example,

Q. Do you know if Jovita was at home that night?

A. Yes.

This simple exchange can indicate either that the deponent knows whether Jovita was at home (the *know* part of the question) or whether Jovita was in fact at home (the *fact* part of the question). While it is not common spoken practice, the questions and answers should go as follows:

Q. Do you know if Jovita was at home that night?

A. Yes, I do know.

Q. Was she at home?

A. No, she was not at home.

A skillful questioner will avoid such compound questions, but alas many are not skillful.

Since the purpose of a deposition is to ascertain what the deponent knows, not to argue the case out of court, the use of argumentative questions is not acceptable. The attorney who accompanies the witness to the deposition will usually object to such a question, and if the opposing attorney wants to know the basis for the objection, the reason is that it is *argumentative*. If one of the attorneys objects to a given question, the witness should refrain from answering until the objection has been resolved. For that reason, it is always advisable for the witness to wait a moment after the end of each question to permit an attorney to raise an objection. If the answer is given too quickly, it stands, in spite of the objection, and may lead to confusion on the record.

Other objections may be made concerning the form of a question, in which case the question is usually reformulated. Often an agreement is reached before the deposition begins that all objections, except for form, will be waived. This simplifies the deposition somewhat. There may be occasions, however, when opposing lawyers will disagree on the propriety of a question and will suspend the proceeding until they can go before a judge to get a ruling. This can become acrimonious and delays the deposition. It is always unfortunate for the deponent when opposing lawyers get into a squabble between themselves, but it does happen.

If the matter is particularly complicated, a deposition may last several hours. It is usually agreed beforehand that if the deponent wishes to take a break during the proceeding, he or she has only to request it. There should be no conversation between the deponent and her or his counsel during the break and no talk between the deponent and other witnesses present in the room.

After the witness has answered all the adversaries' questions, her or his own counsel may wish to ask a few questions to clarify items. Eventually all the questions get asked and answered, some more than once. The deposition is then terminated, with the recorder noting the exact time that the session ended. The transcript of the deposition is prepared and submitted to the interested parties and to the deponent, if desired. Interested parties often waive their privilege of reading, correcting, and signing their deposition records, but deponents should usually insist on reading, making suitable corrections, and signing the document.

If the matter comes to trial, depositions may be (and often are) introduced as supporting evidence and presented to the court. Sections may be read to the court as an indication of previous statements made by the witness, particularly if the witness appears to be changing previous statements in any way. Any discrepancies must be resolved at once and reasons for the changes established.

Investigations and Reports

When litigation has been started and the issues have been defined by the discovery process, it is common to have various investigations done to obtain necessary technical or personal background information. Most law firms have private investigators whom they hire on an as-needed basis to interview various interested parties. They are not the gumshoes commonly portrayed in film productions, but are licensed persons whose mission is to find out certain facts which may or may not be in dispute. They prepare and submit their reports to their clients, setting forth the facts and often commenting on the appearance of the person being interviewed and evaluating how good a witness that person might be if called to testify.

It is important to note that such interviews are not sworn testimony, hence cannot be introduced at trial. A further limitation is that the investigators are not considered to be experts and so cannot give opinions regarding the accuracy or validity of the information they obtain. One of their most important functions is to establish the scenario of what actually took place at the time of the incident which caused the suit to be brought.

Many such investigations take place quite soon after the time of the incident. In addition to talking to witnesses, an essential part of any onsite investigation is the taking of suitable photographs to document important details. Careful notes must be taken for each picture, elucidating what is shown, where the picture was taken, where the photographer was standing, and the direction in which the camera was pointing. Unambiguous identification must be attached so that each picture is positively correlated to what is shown and cannot be mixed up with another picture. In this way a distinct record is obtained which can be preserved for future reference.

In contrast to investigators' factual reports, experts' reports include the interpretation and evaluation of reported facts plus the formulation and submission of conclusions and opinions. To reach the conclusions and opinions, experts collect and evaluate other background information beyond that reported by the investigators. This background information frequently includes transcripts of any sworn testimony which has been taken, interrogatories between the various parties, applicable standards and specifications, regulations promulgated by various government agencies, descriptive information from manufacturers' catalogs, medical records, accident reports by the local police department, or fire incident reports by the fire department.

If physical evidence is involved, in addition to the documentation, experts make a detailed examination of whatever items are available. Photographs of the evidence are taken to document the experts' observations and to identify the particular objects involved. Appropriate measurements are taken; these may involve distances, temperatures, pressures, forces, elevations, angles, light intensity, coefficients of friction, and so forth. Sometimes electrical properties must be examined such as continuity, resistance, voltage, or frequency. All this requires that the expert become thoroughly familiar with the facts of the case, the scenario of what happened, and the background, before performing her or his examination. Only in that way can the important variables be established beforehand and the required instrumentation be made available. Thorough preparation is mandatory before the examination of physical evidence is begun.

One of the frequent, and most stringent, limitations applicable to the examination of physical evidence is that there shall be no damage done to the evidence. Absolutely no destructive testing may be done when that limitation is applied. The obvious reason for this is to preserve the evidence for all interested parties to examine in their turn, but it can result in examinations which are superficial and not really definitive.

The common way to get around this limitation is to arrange for all interested parties to be present when the examination is conducted. Then everyone gets to see the same revelations at the same time. Because of the adversarial nature of litigation, however, it is common that none of the participating investigators makes any audible comment on what is observed because the other parties may hear what is said and use it to their advantage. Great caution must be exercised by the investigator under these circumstances to avoid the accidental disclosure of important information.

A much more relaxed atmosphere prevails when the examiner can work without the other parties being present. Fortunately, this is a common situation. If permission for destructive examination is given, the physical evidence can be taken apart, samples taken for other examinations, and a thorough investigation performed. Usually, the evidence which remains, damaged though it may be, is preserved for possible future reference or examination by others.

The range of physical tests which may be performed is extremely broad. In many cases which involve metal parts, tests are run to establish the hardness of the material, or its tensile strength. If there is a question regarding the composition of the material, chemical analysis may be performed. Metallographic examination may also be carried out to reveal the crystal structure and thereby establish whether the heat treatment of the metal was proper. There are many detailed rules of protocol for the performance of such tests set forth in the standards published by ASTM. Close adherence to the requirements of the standards is mandatory if the reported results are to be accepted as authentic.

Chemical tests likewise involve detailed procedures. With the instruments available today, tests for the presence of certain elements or compounds can be made with extreme accuracy. Parts per billion of certain alleged contaminants can be detected routinely, and both infrared spectroscopy and gas chromatography can be used with incredibly precise results to identify unknown constituents.

In cases where parts of evidence have been broken in service, the surface of the fracture can be examined in extremely fine detail through the use of the scanning electron microscope. In this way the locus of the origin of the fracture can often be identified and associated with some preexisting condition, which enables the expert to establish a causative relationship. The chemical composition of the surface can also be determined for comparison with the specified call-out.

Most of the physical quantities previously mentioned, such as distances, temperatures, light intensity, and voltages, can be measured accurately with commonly available instruments. Most are easily portable and can be carried to the scene by the examiner. One variable in particular, however, falls outside this category. That is the measurement of the *coefficient of friction,* commonly abbreviated as COF. As most engineering textbooks state, the value of the coefficient of friction depends only on the nature of the two rubbing surfaces and is supposed to be independent of the pressure between them and the velocity of their relative motion.

Measurements of the coefficient of friction in the field immediately show that there are practical limitations to the theory. It is extremely difficult, and some would say impossible, to get precise readings for the values of friction coefficients under field conditions. Even under controlled laboratory conditions there is frequently a wide variation in observed values, and the data will have a large "scatter."

Several instruments have been devised to measure the coefficient of friction. Among these are the James machine, the Brungraber machine, and the horizontal-pull slip meter. The James machine is primarily a laboratory instrument, and it is widely used for testing floor surfaces such as vinyl tile. The Brungraber machine, while bulky and heavy to transport, can be used in field settings but only on relatively level surfaces. The horizontal-pull slip meter is easily portable, can be used in moderately inclined situations, but depends to a great extent on the skill of the operator. It has a distinct advantage in that it can be used in close quarters such as on the treads of stairways.

In all cases, several measurements of the coefficient of friction must be taken and a statistical analysis performed on the data to get the best and most reliable value for reporting purposes. The significance of the coefficient of friction becomes very large in slip and fall cases, where it is alleged that the walking surface was unsafe because it was slippery.

Slipperiness has been investigated by many observers over a long time, and the literature is rich with detailed information on precisely how people place their feet during walking. The consensus is that the most dangerous part of the stride occurs when the heel is being brought down onto the walking surface.

For that reason, the investigator must try to determine the exact nature of the material used on the heels and soles of the footwear in use at the time of the incident. A piece of this material should be used as the moving surface of the testing machine, and it should be prepared to have the exact nature of the accident material insofar as possible. The nonmoving surface should be the actual site of the mishap.

Even with all the effort that has been expended in the investigations of slipperiness, there is still no general agreement on how large the coefficient of friction should be to ensure that a given surface is slip-resistant. Some testing laboratories have indicated that 0.50 is the lowest acceptable value which includes a sufficient margin of error on the safe side. Note that such terms as *skidproof* and *nonskid* are essentially meaningless when applied to commercial products. Designers are cautioned to be very conservative when choosing a coefficient of friction because of the great uncertainties present.

Formats for reports vary widely among practitioners, but all invariably include an identification of the parties involved, a setting forth of the various allegations, the objects and scope of the investigation, the background information available, the results of any tests or observations made, and a set of conclusions. Obviously, the conclusions should relate to the allegations cited and must be supportable by the background information and the observations. If the matter moves to the deposition or trial stage, it is certain that the legal people will question the investigator regarding the bases for the conclusions presented, the weight given to certain information, and why some information was discounted in reaching the conclusions.

Courtroom Procedure and Testimony

If the various investigators have done their jobs properly, all the pertinent information will be in the hands of the legal counsel at the end of the discovery process. At that time, they should be able to assess the relative merits of their positions and be able to come to some settlement agreement on the issues. This happens in about 95 percent of civil cases involving engineering input, but in about 1 in 20 cases the settlement process fails. This failure is usually related to the amount of monetary damages the defendant is willing to pay to the plaintiff, but other matters are sometimes at issue as well.

The decision to go to trial entails risk on both sides. The plaintiff is gambling that the trial will result in an award greater than the last offer, with the risk that the award may be less or even nothing. The defendant is betting the other way, and the court is there to settle the differences.

When a person is called as a witness in a trial, either for facts or as an expert, any testimony is given under oath. The actual wording of the oath may vary from one court to another, but absolute truthfulness is demanded in all instances. The penalties for perjury are applicable and are severe. The oath is administered by a court officer, who may or may not use a Bible in the process. After being sworn, the witness is usually asked to state his or her name, spelling the last name. The court reporter records all this and all subsequent words of all concerned parties.

Unlike the deposition process, court testimony is given before a judge and usually (although not always) before a jury. The objective of the trial process is to reach a decision, not just to find out the facts in the matter, as was the case during the discovery process. Arguments between the attorneys take place routinely, and the judge is often called upon to decide between them on the spot.

The witness is first asked a series of questions by the legal representative from the side the witness represents. This is called the *direct examination*. Any facts or expert opinions which the witness may have are brought out during this process. This is "friendly" questioning. Following the direct examination, the opposing attorney begins cross-examination of the witness. This is an "unfriendly" process. Quite literally, the opposing lawyer tries to destroy the credibility of the testimony given by the witness on direct examination. This is the essence of the adversarial process used in the U.S. court system. The jury hears all this and must weigh the value of the given testimony in light of the ability of the witness to substantiate what he or she has said. Credibility is the key factor in this process. If the jury does not believe the witness, no matter how truthful her or his testimony may seem, they will give little or no weight to what has been said.

Courtroom settings vary considerably from place to place. In some municipal courts the room is small, and there is no difficulty in seeing and hearing everything that takes place. In larger courtrooms, however, the size of the room and the often terrible acoustics make communication troublesome. The witness must be careful to speak very clearly so that everyone on the jury can hear what is being said. In contrast to the deposition procedure, models, mock-ups, and other visual aids may be used if the judge permits. A chalkboard is often used. All this is intended to make the testimony of the witness more believable by the jury.

Following the direct examination and the cross-examination, there may be some additional questions which the lawyer working with the witness will wish to ask. This is known as a *redirect examination,* and frequently it involves an attempt to restore credibility to some testimony which was attacked by the adversary. Such efforts are known as *rehabilitation* of the witness. Some new information may also be brought out, and this is subject to cross-examination again, just as the original direct examination was. When none of the attorneys has any more questions, the witness is excused and steps down from the witness stand.

The witness should understand that the judge is the absolute ruler in the courtroom. When the judge enters the room, it is common for the bailiff to announce the fact and to call upon all persons present to rise and to remain standing until the judge sits down. This deference to the judge's position continues throughout the trial process. When the judge is addressed, it is always as "your Honor," and a specific request is made by the attorneys for permission to approach the bench or a witness before they attempt to do so. Such rigid decorum may seem strange to a first-time witness, but the protocols must be observed carefully to avoid censure by the court.

In addition to the judge, court officers are present. A clerk, a bailiff, and a recorder are there to assist the judge in the performance of her or his duties. The bailiff escorts the members of the jury into and out of the courtroom, making certain that they have no contact with outsiders in the process. A witness must be extremely careful not to speak to any jury members encountered outside the courtroom, not even to say hello or to pass the time of day. While this may seem to be unfriendly and almost antisocial, any communication with the jury outside the courtroom is subject to the interpretation that the witness was exercising undue influence. To be absolutely certain that the trial is a fair one, all such communications are absolutely forbidden. If there is a breach of this restriction, a mistrial may be declared because of the witness's indiscretion.

The recorder performs the same function in the court as during a deposition. Every spoken word is taken down and included in the transcript of the proceedings. As noted previously, the recorder cannot take down gestures, smiles, shakes or nods of the head, or other nonaudible responses. The witness is required to observe these limitations and will be bluntly reminded if any lapses occur.

When exhibits are presented to the court, the judge will have them identified and marked by the recorder. Only after due consideration are exhibits admitted as evidence in the case, and such admissibility is not always granted by the judge. Keeping possibly adverse evidence out of the proceedings is one thing lawyers attempt to do, but the judge makes the final decision, yes or no.

As the referee in the proceedings, the judge is often called upon to make decisions regarding the propriety of certain questions addressed to the witness. One lawyer may object to a question, based on its form or its content. If this happens, the witness should refrain from replying to the question until the objection has been ruled on by the judge. If the objection is sustained, the question must not be answered. To allow for an objection, it is always advisable for the witness to wait a few seconds after the end of a question before replying.

Sometimes while a witness is on the stand, a legal dispute may arise between the opposing lawyers, and they will request a consultation with the judge. This is called a *sidebar conference,* and it is conducted in hushed voices so that the content of the discussion cannot be overheard by the jury or by the witness. The recorder is present, however, and takes down whatever is said as part of the record.

Some courts permit witnesses to be present during the testimony of other witnesses, and others prohibit it. When it is permitted, it is particularly advantageous to defense witnesses to be present during the presentation of the testimony of plaintiff's witnesses. Having heard the arguments for the plaintiff, a defense witness is in a position to refute such testimony. The defense witness's attorney may ask questions, based on the previous testimony, knowing that it has been heard by the witness. Unfortunately, plaintiff's witnesses cannot enjoy this advantage since they always testify first. Obviously, the situation cannot arise in courts where no witnesses are permitted to hear the testimony of others.

It is common for the attorney representing the witness's side to have a conference with the witness prior to putting her or him on the stand. One thing every lawyer learns early is to never ask a question whose answer is uncertain. The pretestimony conference is held to make certain that there are no surprises in what the witness is going to say. If some troublesome areas are revealed during the conference, the attorney may decide to avoid them during the direct examination of the witness. Under no circumstances should the witness volunteer information which is adverse to the party they are representing in order to tell "the whole truth" about a particular item. Remember that the attorney is acting in the role of an advocate for his or her principal and presents only information which is favorable to the cause. It is the task of the opposing attorney to bring out anything which is unfavorable. Under the U.S. adversarial system, the trial is a conflict that both sides are trying to win. For an engineer, trained in the laws of nature, it may be troublesome to watch the way some members of the legal profession handle the facts. However, the jury, not the witness, is there to decide which version of the facts is true.

The pretrial conference is also an opportunity for the attorney to ask technical questions of the witness to find out what to ask the opposing witnesses under cross-examination. Having read the witness's reports along with the depositions taken during the discovery process, the lawyer is familiar with the issues involved and their legal aspects, but may not be very well informed or have more than superficial knowledge of the nuances of the technical aspects. It is at this juncture that the qualified technical expert can be of great assistance to the lawyer by pointing out where the opposing case is weakest, what its strengths are, what to attack, and what to avoid. This strategy session often includes the parties to the action, when they are available.

Experts and Opinions

As pointed out previously, nonexpert witnesses can only testify regarding the facts of the matter as they know them from first-hand, direct experience. They are not considered to be qualified to give opinions or interpretations of those facts. Expert witnesses, however, can and do give interpretations and opinions when requested to do so. Before the requests are made, however, the attorney working with the expert must prove to the court that the witness is indeed an expert. After the witness has been seated on the witness stand and identified himself or herself to the court, the witness's attorney will ask the witness to describe in detail her or his educational background, degrees obtained, field of specialization, and any special courses or seminars attended. After the educational data are given, the witness's work history is requested, along with any professional society memberships and offices held in those organizations. If the witness has written any publications, those are described as they relate to the matter at hand.

After putting these qualifications on record, the lawyer for the side the witness is testifying for will offer the witness to the court as an expert. The opposing lawyer then has the opportunity to question the proffered witness about the details and relevance of her or his background and experience. If this lawyer was present at any depositions previously given by this witness, the lawyer will have had time to check into the witness's background to find any weaknesses or places where the expertise is thin. These facts are brought to the attention of the court in the presence of the jury, so that they can give proper credence to the witness and her or his testimony. At the end of this questioning, the opposing attorney may request that the proffered witness not be accepted as an expert. The judge must then make a decision of whether to accept the witness and, if the witness is accepted, whether to place any limitations on the fields of expertise on which the witness can offer opinions.

After the witness has been qualified, the questioning turns to the facts of the case at hand. Any reports submitted by the expert in the course of his or her investigations are usually introduced, first as exhibits and later as evidence for the jury to consider in its deliberations. Detailed questions about statements in the reports are asked by attorneys from both sides, and so it is imperative that the witness review such reports very carefully prior to taking the stand to defend them.

Eventually the witness's attorney asks what is called the *hypothetical* question. A more or less lengthy list of assumptions is given to the witness. The assumptions should be based on facts already brought out in the testimony of prior witnesses, if the hypothetical question is to have any real validity. Once the hypothetical question is formulated, the attorney will ask,

> Q. Based on these assumed facts, do you have an opinion regarding the causation of the alleged incident?

Note that the question is only, Do you have an opinion? *not* What is your opinion? The proper reply to this question is

> A. Yes, I have an opinion.

The attorney will then ask,

> Q. What is your opinion?

The witness will then, and only then, give an opinion.

Of course, the opposing attorney will challenge this opinion and attempt to destroy its validity on cross-examination of the witness. Routinely the attack includes the introduction of as many additional factors as the challenger can find which were not specifically addressed by the witness. This is an attempt to question the thoroughness of the investigation. A standard question at this juncture is to ask if the additional factor was considered, and if not, why not. This is usually followed by, If you had considered it, would that factor have changed your opinion?

As any experienced engineer knows, the possible avenues of investigation can be very large, even for seemingly trivial cases. If all the possible factors are examined in complete detail, the expense of carrying out the investigation can be totally unreasonable. The professional engineer is obligated to recognize the possible causative factors in an incident in her or his preliminary assessment of the matter and then to evaluate each on a cost-effectiveness basis to decide whether it is worth the effort to pursue. A reply to the effect that such an analysis and evaluation were carried out is usually sufficient to defuse the opposition's attack. This implies that such an effort was actually performed. If it was not, the reply is not truthful, and the witness is exposed to the hazard of being impeached. If that happens, the witness is totally destroyed, and the opinions given are considered worthless.

Summary

To engineers and other technically or scientifically trained persons, the legal process is a totally different world from what they are accustomed to. Under the U.S. judicial system, the legal process for the correction of some alleged wrongful act is a contest between adversaries. The contest is called a civil trial. The objective of this contest is to win. It is a confrontation carried out under strict rules of decorum and procedure which have been established by law and by longstanding custom.

Obviously, the best way to stay out of this confrontational situation is to avoid committing any wrongful acts of omission or commission. That is, do the design job correctly in the first place.

Appendix A

The Consumer Product Safety Act of 1972

CONSUMER PRODUCT SAFETY ACT

(PUBLIC LAW 92–573; OCT. 27, 1972)

SHORT TITLE; TABLE OF CONTENTS

SECTION 1. This Act may be cited as the "Consumer Product Safety Act".

15 U.S.C. 2051 note

TABLE OF CONTENTS

FINDINGS AND PURPOSES

SEC. 2. (a) The Congress finds that—

15 U.S.C. 2051

(1) an unacceptable number of consumer products which present unreasonable risks of injury are distributed in commerce;

(2) complexities of consumer products and the diverse nature and abilities of consumers using them

frequently result in an inability of users to anticipate risks and to safeguard themselves adequately;

(3) the public should be protected against unreasonable risks of injury associated with consumer products;

(4) control by State and local governments of unreasonable risks of injury associated with consumer products is inadequate and may be burdensome to manufacturers;

(5) existing Federal authority to protect consumers from exposure to consumer products presenting unreasonable risks of injury is inadequate; and

(6) regulation of consumer products the distribution or use of which affects interstate or foreign commerce is necessary to carry out this Act.

(b) The purposes of this Act are—

(1) to protect the public against unreasonable risks of injury associated with consumer products;

(2) to assist consumers in evaluating the comparative safety of consumer products;

(3) to develop uniform safety standards for consumer products and to minimize conflicting State and local regulations; and

(4) to promote research and investigation into the causes and prevention of product-related deaths, illnesses, and injuries.

DEFINITIONS

15 U.S.C. 2052

SEC. 3. (a) For purposes of this Act:

(1) The term "consumer product" means any article, or component part thereof, produced or distributed (i) for sale to a consumer for use in or around a permanent or temporary household or residence, a school, in recreation, or otherwise, or (ii) for the personal use, consumption or enjoyment of a consumer in or around a permanent or temporary household or residence, a school, in recreation, or otherwise; but such term does not include—

(A) any article which is not customarily produced or distributed for sale to, or use or consumption by, or enjoyment of, a consumer,

(B) tobacco and tobacco products,

(C) motor vehicles or motor vehicle equipment (as defined by sections 102 (3) and (4) of the National Traffic and Motor Vehicle Safety Act of 1966),

(D) pesticides (as defined by the Federal Insecticide, Fungicide, and Rodenticide Act),

(E) any article which, if sold by the manufacturer, producer, or importer, would be subject to the tax imposed by section 4181 of the Internal Revenue Code of 1954 (determined without regard to any exemptions from such tax provided by section 4182 or 4221, or any other provision of such Code), or any component of any such article,

(F) aircraft, aircraft engines, propellers, or appliances (as defined in section 101 of the Federal Aviation Act of 1958),

(G) boats which could be subjected to safety regulation under the Federal Boat Safety Act of 1971 (46 U.S.C. 1451 et seq.); vessels, and appurtenances to vessels (other than such boats), which could be subjected to safety regulation under title 52 of the Revised Statutes or other marine safety statutes administered by the department in which the Coast Guard is operating; and equipment (including associated equipment, as defined in section 3(8) of the Federal Boat Safety Act of 1971) to the extent that a risk of injury associated with the use of such equipment on boats or vessels could be eliminated or reduced by action taken under any statute referred to in this subparagraph,

(H) drugs, devices, or cosmetics (as such terms are defined in sections 201 (g), (h), and (i) of the Federal Food, Drug, and Cosmetic Act), or

(I) food. The term "food", as used in this subparagraph means all "food", as defined in section 201(f) of the Federal Food, Drug, and Cosmetic Act, including poultry and poultry products (as defined in sections 4 (e) and (f) of the Poultry Products Inspection Act), meat, meat food products (as defined in section 1(j) of the Federal Meat Inspection Act), and eggs and egg products (as defined in section 4 of the Egg Products Inspection Act).

Except for the regulation under this Act or the Federal Hazardous Substances Act of fireworks devices or any substance intended for use as a component of any such device, the Commission shall have no authority under the functions transferred pursuant to section 30 of this Act to regulate any product or article described in subparagraph (E) of this paragraph or described, without regard to quantity, in section 845(a)(5) of title 18, United States Code. See sections 30(d) and 31 of this Act, for other limitations on Commission's authority to regulate certain consumer products.

(2) The term "consumer product safety rule" means a consumer products safety standard described

in section 7(a), or a rule under this Act declaring a consumer product a banned hazardous product.

(3) The term "risk of injury" means a risk of death, personal injury, or serious or frequent illness.

(4) The term "manufacturer" means any person who manufactures or imports a consumer product.

(5) The term "distributor" means a person to whom a consumer product is delivered or sold for purposes of distribution in commerce, except that such term does not include a manufacturer or retailer of such product.

(6) The term "retailer" means a person to whom a consumer product is delivered or sold for purposes of sale or distribution by such person to a consumer.

(7)(A) The term "private labeler" means an owner of a brand or trademark on the label of a consumer product which bears a private label.

(B) A consumer product bears a private label if (i) the product (or its container) is labeled with the brand or trademark of a person other than a manufacturer of the product, (ii) the person with whose brand or trademark the product (or container) is labeled has authorized or caused the product to be so labeled, and (iii) the brand or trademark of a manufacturer of such product does not appear on such label.

(8) The term "manufactured" means to manufacture, produce, or assemble.

(9) The term "Commission" means the Consumer Product Safety Commission, established by section 4.

(10) The term "State" means a State, the District of Columbia, the Commonwealth of Puerto Rico, the Virgin Islands, Guam, Wake Island, Midway Island, Kingman Reef, Johnston Island, the Canal Zone, American Samoa, or the Trust Territory of the Pacific Islands.

(11) The terms "to distribute in commerce" and "distribution in commerce" means to sell in commerce, to introduce or deliver for introduction into commerce, or to hold for sale or distribution after introduction into commerce.

(12) The term "commerce" means trade, traffic, commerce, or transportation—

(A) between a place in a State and any place outside thereof, or

(B) which affects trade, traffic, commerce, or transportation described in subparagraph (A).

(13) The terms "import" and "importation" include reimporting a consumer product manufactured or processed, in whole or in part, in the United States.

(14) The term "United States", when used in the geographic sense, means all of the States (as defined in paragraph (10)).

(b) A common carrier, contract carrier, or freight forwarder shall not, for purposes of this Act, be deemed to be a manufactuer, distributor, or retailer of a consumer product solely by reason of receiving or transporting a consumer product in the ordinary course of its business as such a carrier or forwarder.

CONSUMER PRODUCT SAFETY COMMISSION

15 U.S.C. 2053

SEC. 4. (a) An independent regulatory commission is hereby established, to be known as the Consumer Product Safety Commission, consisting of five Commissioners who shall be appointed by the President, by and with the advice and consent of the Senate, one of whom shall be designated by the President as Chairman. The Chairman, when so designated shall act as Chairman until the expiration of his term of office as Commissioner. Any member of the Commission may be removed by the President for neglect of duty or malfeasance in office but for no other cause.

(b)(1) Except as provided in paragraph (2), (A) the Commissioners first appointed under this section shall be appointed for terms ending three, four, five, six, and seven years, respectively, after the date of the enactment of this Act, the term of each to be designated by the President at the time of nomination; and (B) each of their successors shall be appointed for a term of seven years from the date of the expiration of the term for which his predecessor was appointed.

(2) Any Commissioner appointed to fill a vacancy occurring prior to the expiration of the term for which his predecessor was appointed shall be appointed only for the remainder of such term. A Commissioner may continue to serve after the expiration of his term until his successor has taken office, except that he may not so continue to serve more than one year after the date on which his term would otherwise expire under this subsection.

(c) Not more than three of the Commissioners shall be affiliated with the same political party. No individual (1) in the employ of, or holding any official relation to, any person engaged in selling or manufacturing consumer products, or (2) owning stock or bonds of substantial value in a person so engaged, or (3) who is in any other manner pecuniarily interested in such a person, or in a substantial supplier of such a person, shall hold the office of Commissioner. A Commissioner may not engage in any other business, vocation, or employment.

(d) No vacancy in the Commission shall impair the right of the remaining Commissioners to exercise all the powers of the Commission, but three members of the

Commission shall constitute a quorum for the transaction of business. The Commission shall have an official seal of which judicial notice shall be taken. The Commission shall annually elect a Vice Chairman to act in the absence or disability of the Chairman or in case of a vacancy in the office of the Chairman.

(e) The Commission shall maintain a principal office and such field offices as it deems necessary and may meet and exercise any of its powers at any other place.

(f)(1) The Chairman of the Commission shall be the principal executive officer of the Commission, and he shall exercise all of the executive and administrative functions of the Commission, including functions of the Commission with respect to (A) the appointment and supervision of personnel employed under the Commission (other than personnel employed regularly and full time in the immediate offices of commissioners other than the Chairman), (B) the distribution of business among personnel appointed and supervised by the Chairman and among administrative units of the Commission, and (C) the use and expenditure of funds.

(2) In carrying out any of his functions under the provisions of this subsection the Chairman shall be governed by general policies of the Commission and by such regulatory decisions, findings, and determinations as the Commission may by law be authorized to make.

(3) Requests or estimates for regular, supplemental, or deficiency appropriations on behalf of the Commission may not be submitted by the Chairman without the prior approval of the Commission.

(g)(1) The Chairman, subject to the approval of the Commission, shall appoint an Executive Director, a General Counsel, a Director of Engineering Sciences, a Director of Epidemiology, and a Director of Information. No individual so appointed may receive pay in excess of the annual rate of basic pay in effect for grade GS–18 of the General Schedule.

(2) The Chairman, subject to subsection (f)(2), may employ such other officers and employees (including attorneys) as are necessary in the execution of the Commission's functions. No regular officer or employee of the Commission who was at any time during the 12 months preceding the termination of his employment with the Commission compensated at a rate in excess of the annual rate of basic pay in effect for grade GS–14 of the General Schedule, shall accept employment or compensation from any manufacturer subject to this Act, for a period of 12 months after terminating employment with the Commission.

(3) In addition to the number of positions authorized by section 5108(a) of title 5, United States Code, the Chairman, subject to the approval of the Commission,

and subject to the standards and procedures prescribed by chapter 51 of title 5, United States Code, may place a total of twelve positions in grades GS-16, GS-17, and GS-18.

(4) The appointment of any officer (other than a Commissioner) or employee of the Commission shall not be subject, directly or indirectly, to review or approval by any officer or entity within the Executive Office of the President.

(h)(1) Section 5314 of title 5, United States Code, is amended by adding at the end thereof the following new paragraph:

"(59) Chairman, Consumer Product Safety Commission."

(2) Section 5315 of such title is amended by adding at the end thereof the following new paragraph:

"(97) Members, Consumer Product Safety Commission (4)."

(i) Subsections (a) and (h) of section 2680 of title 28, United States Code, do not prohibit the bringing of a civil action on a claim against the United States which—

(1) is based upon—

(A) misrepresentation or deceit before January 1, 1978, on the part of the Commission or any employee thereof, or

(B) any exercise or performance, or failure to exercise or perform, a discretionary function on the part of the Commission or any employee thereof before January 1, 1978, which exercise, performance, or failure was grossly negligent; and

(2) is not made with respect to any agency action (as defined in section 551(13) of title 5, United States Code).

In the case of a civil action on a claim based upon the exercise or performance of, or failure to exercise or perform, a discretionary function, no judgment may be entered against the United States unless the court in which such action was brought determines (based upon consideration of all the relevant circumstances, including the statutory responsibility of the Commission and the public interest in encouraging rather than inhibiting the exercise of discretion) that such exercise, performance, or failure to exercise or perform was unreasonable

PRODUCT SAFETY INFORMATION AND RESEARCH

SEC. 5. (a) The Commission shall—

(1) maintain an Injury Information Clearinghouse to collect, investigate, analyze, and disseminate injury data, and information, relating to the

causes and prevention of death, injury, and illness associated with consumer products; and

(2) conduct such continuing studies and investigations of deaths, injuries, diseases, other health impairments, and economic losses resulting from accidents involving consumer products as it deems necessary.

(b) The Commission may—

(1) conduct research, studies, and investigations on the safety of consumer products and on improving the safety of such products;

(2) test consumer products and develop product safety test methods and testing devices; and

(3) offer training in product safety investigation and test methods, and assist public and private organizations, administratively and technically, in the development of safety standards and test methods.

(c) In carrying out its functions under this section, the Commission may make grants or enter into contracts for the conduct of such functions with any person (including a governmental entity).

(d) Whenever the Federal contribution for any information, research, or development activity authorized by this Act is more than minimal, the Commission shall include in any contract, grant, or other arrangement for such activity, provisions effective to insure that the rights to all information, uses, processes, patents, and other developments resulting from that activity will be made available to the public without charge on a nonexclusive basis. Nothing in this subsection shall be construed to deprive any person of any right which he may have had, prior to entering into any arrangement referred to in this subsection, to any patent, patent application, or invention.

PUBLIC DISCLOSURE OF INFORMATION

15 U.S.C. 2055

SEC. 6. (a)(1) Nothing contained in this Act shall be deemed to require the release of any information described by subsection (b) of section 552, title 5, United States Code, or which is otherwise protected by law from disclosure to the public.

(2) All information reported to or otherwise obtained by the Commission or its representative under this Act which information contains or relates to a trade secret or other matter referred to in section 1905 of title 18, United States Code, shall be considered confidential and shall not be disclosed, except that such information may be disclosed to other officers or employees concerned with carrying out this Act or when relevant in any proceeding under this Act. Nothing in this Act shall authorize the withholding of information by the Commission or any

officer or employee under its control from the duly authorized committees of the Congress.

(b)(1) Except as provided by paragraph (2) of this subsection, not less than 30 days prior to its public disclosure of any information obtained under this Act, or to be disclosed to the public in connection therewith (unless the Commission finds out that the public health and safety requires a lesser period of notice), the Commission shall, to the extent practicable, notify, and provide a summary of the information to, each manufacturer or private labeler of any consumer product to which such information pertains, if the manner in which such consumer product is to be designated or described in such information will permit the public to ascertain readily the identity of such manufacturer or private labeler, and shall provide such manufacturer or private labeler with a reasonable opportunity to submit comments to the Commission in regard to such information. The Commission shall take reasonable steps to assure, prior to its public disclosure thereof, that information from which the identity of such manufacturer or private labeler may be readily ascertained is accurate, and that such disclosure is fair in the circumstances and reasonably related to effectuating the purposes of this Act. If the Commission finds that, in the administration of this Act, it has made public disclosure of inaccurate or misleading information which reflects adversely upon the safety of any consumer product, or the practices of any manufacturer, private labeler, distributor, or retailer of consumer products, it shall, in a manner similar to that in which such disclosure was made, publish a retraction of such inaccurate or misleading information.

(2) Paragraph (1) (except for the last sentence thereof) shall not apply to the public disclosure of (A) information about any consumer product with respect to which product the Commission has filed an action under section 12 (relating to imminently hazardous products), or which the Commission has reasonable cause to believe is in violation of section 19 (relating to prohibited acts), or (B) information in the course of or concerning any administrative or judicial proceeding under this Act.

(c) The Commission shall communicate to each manufacturer of a consumer product, insofar as may be practicable, information as to any significant risk of injury associated with such product.

CONSUMER PRODUCT SAFETY STANDARDS

SEC. 7. (a)(1) The Commission may by rule, in accordance with this section and section 9, promulgate con-

15 U.S.C. 2056

sumer product safety standards. A consumer product safety standard shall consist of one or more of any of the following types of requirements:

(A) Requirements as to performance, composition, contents, design, construction, finish, or packaging of a consumer product.

(B) Requirements that a consumer product be marked with or accompanied by clear and adequate warnings or instructions, or requirements respecting the form of warnings or instructions.

Any requirement of such a standard shall be reasonably necessary to prevent or reduce an unreasonable risk of injury associated with such product. The requirements of such a standard (other than requirements relating to labeling, warnings, or instructions) shall, whenever feasible, be expressed in terms of performance requirements.

(2) No consumer product safety standard promulgated under this section shall require, incorporate, or reference any sampling plan. The preceding sentence shall not apply with respect to any consumer product safety standard or other agency action of the Commission under this Act (A) applicable to a fabric, related material, or product which is subject to a flammability standard or for which a flammability standard or other regulation may be promulgated under the Flammable Fabrics Act, or (B) which is or may be applicable to glass containers.

(b) A proceeding for the development of a consumer product safety standard under this Act shall be commenced by the publication in the Federal Register of a notice which shall—

(1) identify the product and the nature of the risk of injury associated with the product;

(2) state the Commission's determination that a consumer product safety standard is necessary to eliminate or reduce the risk of injury;

(3) include information with respect to any existing standard known to the Commission which may be relevant to the proceeding; and

(4) include an invitation for any person, including any State or Federal agency (other than the Commission), within 30 days after the date of publication of the notice (A) to submit to the Commission an existing standard as the proposed consumer product safety standard or (B) to offer to develop the proposed consumer product safety standard.

An invitation under paragraph (4)(B) shall specify the period of time in which the offeror of an accepted offer is to develop the proposed standard. The period specified shall be a period ending 150 days after the date the offer is accepted unless the Commission for good cause finds (and includes such finding in the notice) that a different period is appropriate.

(c) If the Commission determines that (1) there exists a standard which has been issued or adopted by any Federal agency or by any other qualified agency, organization, or institution, and (2) such standard if promulgated under this Act, would eliminate or reduce the unreasonable risk of injury associated with the product, then it may, in lieu of accepting an offer pursuant to subsection (d) of this section, publish such standard as a proposed consumer product safety rule.

(d)(1) Except as provided by subsection (c), the Commission shall accept one, and may accept more than one, offer to develop a proposed consumer product safety standard pursuant to the invitation prescribed by subsection (b)(4)(B), if it determines that the offeror is technically competent, is likely to develop an appropriate standard within the period specified in the invitation under subsection (b), and will comply with regulations of the Commission under paragraph (3) of this subsection. The Commission shall publish in the Federal Register the name and address of each person whose offer it accepts, and a summary of the terms of such offer as accepted.

(2) If an offer is accepted under this subsection, the Commission may agree to contribute to the offeror's cost in developing a proposed consumer product safety standard, in any case in which the Commission determines that such contribution is likely to result in a more satisfactory standard than would be developed without such contribution, and that the offeror is financially responsible. Regulations of the Commission shall set forth the items of cost in which it may participate, and shall exclude any contribution to the acquisition of land or buildings. Payments under agreements entered into under this paragraph may be made without regard to section 3648 of the Revised Statutes of the United States (31 U.S.C. 529).

(3) The Commission shall prescribe regulations governing the development of proposed consumer product safety standards by persons whose offers are accepted under paragraph (1). Such regulations shall include requirements—

(A) that standards recommended for promulgation be suitable for promulgation under this Act, be supported by test data or such other documents or materials as the Commission may reasonably require to be developed, and (where appropriate) contain suitable test methods for measurement of compliance with such standards;

(B) for notice and opportunity by interested persons (including representatives of consumers and consumer organizations) to participate in the development of such standards;

(C) for the maintenance of records, which shall be available to the public, to disclose the course of the development of standards recommended for promulgation, the comments and other information submitted by any person in connection with such development (including dissenting views and comments and information with respect to the need for such recommended standards), and such other matters as may be relevant to the evaluation of such recommended standards; and

(D) that the Commission and the Comptroller General of the United States, or any of their duly authorized representatives, have access for the purpose of audit and examination to any books, documents, papers, and records relevant to the development of such recommended standards or to the expenditure of any contribution of the Commission for the development of such standards.

(e)(1) If the Commission publishes a notice pursuant to subsection (b) to commence a proceeding for the development of a consumer product safety standard for a consumer product and if—

(A) the Commission does not, within 30 days after the date of publication of such notice, accept an offer to develop such a standard, or

(B) the development period (specified in paragraph (3)) for such standard ends,

the Commission may develop a proposed consumer product safety rule respecting such product and publish such proposed rule.

(2) If the Commission accepts an offer to develop a proposed consumer product safety standard, the Commission may not, during the development period (specified in paragraph (3)) for such standard—

(A) publish a proposed rule applicable to the same risk of injury associated with such product, or

(B) develop proposals for such standard or contract with third parties for such development, unless the Commission determines that no offeror whose offer was accepted is making satisfactory progress in the development of such standard.

In any case in which the sole offeror whose offer is accepted under subsection (d)(1) of this section is the manufacturer, distributor, or retailer of a consumer product proposed to be regulated by the consumer product safety standard, the Commission may independently proceed to develop proposals for such standard during the development period.

(3) For purposes of paragraph (2), the development period for any standard is a period (A) beginning on the

date on which the Commission first accepts an offer under subsection (d)(1) for the development of a proposed standard, and (B) ending on the earlier of—

(i) the end of the period specified in the notice of proceeding (except that the period specified in the notice may be extended if good cause is shown and the reasons for such extension are published in the Federal Register), or

(ii) the date on which it determines (in accordance with such procedures as it may by rule prescribe) that no offeror whose offer was accepted is able and willing to continue satisfactorily the development of the proposed standard which was the subject of the offer, or

(iii) the date on which an offeror whose offer was accepted submits such a recommended standard to the Commission.

(f) If the Commission publishes a notice pursuant to subsection (b) to commence a proceeding for the development of a consumer product safety standard and if—

(1) no offer to develop such a standard is submitted to, or, if such an offer is submitted to the Commission, no such offer is accepted by, the Commission within a period of 60 days from the publication of such notice (or within such longer period as the Commission may prescribe by a notice published in the Federal Register stating good cause therefor), the Commission shall—

(A) by notice published in the Federal Register terminate the proceeding begun by the subsection (b) notice, or

(B) develop proposals for a consumer product safety rule for a consumer product identified in the subsection (b) notice and within a period of 150 days (or within such longer period as the Commission may prescribe by a notice published in the Federal Register stating good cause therefor) from the expiration of the 60-day (or longer) period—

(i) by notice published in the Federal Register terminate the proceeding begun by the subsection (b) notice, or

(ii) publish a proposed consumer product safety rule; or

(2) an offer to develop such a standard is submitted to and accepted by the Commission within the 60-day (or longer) period, then not later than 210 days (or such later time as the Commission may prescribe by notice published in the Federal Register stating good cause therefor) after the date of the acceptance of such offer the Commission shall take the action described in clause (i) or (ii) of paragraph (1)(B).

BANNED HAZARDOUS PRODUCTS

15 U.S.C. 2057

SEC. 8. Whenever the Commission finds that—

(1) a consumer product is being, or will be, distributed in commerce and such consumer product presents an unreasonable risk of injury; and

(2) no feasible consumer product safety standard under this Act would adequately protect the public from the unreasonable risk of injury associated with such product,

the Commission may propose and, in accordance with section 9, promulgate a rule declaring such product a banned hazardous product.

ADMINISTRATIVE PROCEDURE APPLICABLE TO PROMULGATION OF CONSUMER PRODUCT SAFETY RULES

15 U.S.C. 2058

SEC. 9. (a)(1) Within 60 days after the publication under section 7 (c), (e)(1), or (f) or section 8 of a proposed consumer product safety rule respecting a risk of injury associated with a consumer product, the Commission shall—

(A) promulgate a consumer product safety rule respecting the risk of injury associated with such product if it makes the findings required under subsection (c), or

(B) withdraw by rule the applicable notice of proceeding if it determines that such rule is not (i) reasonably necessary to eliminate or reduce an unreasonable risk of injury associated with the product, or (ii) in the public interest:

except that the Commission may extend such 60-day period for good cause shown (if it publishes its reasons therefor in the Federal Register).

(2) Consumer product safety rules which have been proposed under section 7 (c), (e)(1), or (f) or section 8 shall be promulgated pursuant to section 553 of title 5, United States Code, except that the Commission shall give interested persons an opportunity for the oral presentation of data, views, or arguments, in addition to an opportunity to make written submissions. A transcript shall be kept of any oral presentation.

(b) A consumer product safety rule shall express in the rule itself the risk of injury which the standard is designed to eliminate or reduce. In promulgating such a rule the Commission shall consider relevant available product data including the results of research, development, testing, and investigation activities conducted generally and pursuant to this Act. In the promulgation of such a rule the Commission shall also consider and take into account the special needs of elderly and handicapped persons to determine the extent to which such persons may be adversely affected by such rule.

(c)(1) Prior to promulgating a consumer product safety rule, the Commission shall consider, and shall make appropriate findings for inclusion in such rule with respect to—

(A) the degree and nature of the risk of injury the rule is designed to eliminate or reduce;

(B) the approximate number of consumer products, or types or classes thereof, subject to such rule;

(C) the need of the public for the consumer products subject to such rule, and the probable effect of such rule upon the utility, cost, or availability of such products to meet such need; and

(D) any means of achieving the objective of the order while minimizing adverse effects on competition or disruption or dislocation of manufacturing and other commercial practices consistent with the public health and safety.

(2) The Commission shall not promulgate a consumer product safety rule unless it finds (and includes such finding in the rule)—

(A) that the rule (including its effective date) is reasonably necessary to eliminate or reduce an unreasonable risk of injury associated with such product;

(B) that the promulgation of the rule is in the public interest; and

(C) in the case of a rule declaring the product a banned hazardous product, that no feasible consumer product safety standard under this Act would adequately protect the public from the unreasonable risk of injury associated with such product.

(d)(1) Each consumer product safety rule shall specify the date such rule is to take effect not exceeding 180 days from the date promulgated, unless the Commission finds, for good cause shown, that a later effective date is in the public interest and publishes its reasons for such finding. The effective date of a consumer product safety standard under this Act shall be set at a date at least 30 days after the date of promulgation unless the Commission for good cause shown determines that an earlier effective date is in the public interest. In no case may the effective date be set at a date which is earlier than the date of promulgation. A consumer product safety standard shall be applicable only to consumer products manufactured after the effective date.

(2) The Commission may by rule prohibit a manufacturer of a consumer product from stockpiling any product to which a consumer product safety rule applies, so as to prevent such manufacturer from circumventing

the purpose of such consumer product safety rule. For purposes of this paragraph, the term "stockpiling" means manufacturing or importing a product between the date of promulgation of such consumer product safety rule and its effective date at a rate which is significantly greater (as determined under the rule under this paragraph) than the rate at which such product was produced or imported during a base period (prescribed in the rule under this paragraph) ending before the date of promulgation of the consumer product safety rule.

(e) The Commission may by rule amend or revoke any consumer product safety rule. Such amendment or revocation shall specify the date on which it is to take effect which shall not exceed 180 days from the date the amendment or revocation is published unless the Commission finds for good cause shown that a later effective date is in the public interest and publishes its reasons for such finding. Where an amendment involves a material change in a consumer product safety rule, sections 7 and 8, and subsections (a) through (d) of this section shall apply. In order to revoke a consumer product safety rule, the Commission shall publish a proposal to revoke such rule in the Federal Register, and allow oral and written presentations in accordance with subsection (a)(2) of this section. It may revoke such rule only if it determines that the rule is not reasonably necessary to eliminate or reduce an unreasonable risk of injury associated with the product. Section 11 shall apply to any amendment of a consumer product safety rule which involves a material change and to any revocation of a consumer product safety rule, in the same manner and to the same extent as such section applies to the Commission's action in promulgating such a rule.

COMMISSION RESPONSIBILITY—PETITION FOR CONSUMER PRODUCT SAFETY RULE

15 U.S.C. 2059

SEC. 10. (a) Any interested person, including a consumer or consumer organization, may petition the Commission to commence a proceeding for the issuance, amendment, or revocation of a consumer product safety rule.

(b) Such petition shall be filed in the principal office of the Commission and shall set forth (1) facts which it is claimed establish that a consumer product safety rule or an amendment or revocation thereof is necessary, and (2) a brief description of the substance of the consumer product safety rule or amendment thereof which it is claimed should be issued by the Commission.

(c) The Commission may hold a public hearing or may conduct such investigation or proceeding as it deems appropriate in order to determine whether or not such petition should be granted.

(d) Within 120 days after filing of a petition described in subsection (b), the Commission shall either grant or deny the petition. If the Commission grants such petition, it shall promptly commence an appropriate proceeding under section 7 or 8. If the Commission denies such petition it shall publish in the Federal Register its reasons for such denial.

(e)(1) If the Commission denies a petition made under this section (or if it fails to grant or deny such petition within the 120-day period) the petitioner may commence a civil action in a United States district court to compel the Commission to initiate a proceeding to take the action requested. Any such action shall be filed within 60 days after the Commission's denial of the petition, or (if the Commission fails to grant or deny the petition within 120 days after filing the petition) within 60 days after the expiration of the 120-day period.

(2) If the petitioner can demonstrate to the satisfaction of the court, by a preponderance of evidence in a de novo proceeding before such court, that the consumer product presents an unreasonable risk of injury, and that the failure of the Commission to initiate a rulemaking proceeding under section 7 or 8 unreasonably exposes the petitioner or other consumers to a risk of injury pesented by the consumer product, the court shall order the Commission to initiate the action requested by the petitioner.

(3) In any action under this subsection, the district court shall have no authority to compel the Commission to take any action other than the initiation of a rulemaking proceeding in accordance with section 7 or 8.

(4) In any action under this subsection the court may in the interest of justice award the costs of suit, including reasonable attorneys' fees and reasonable expert witnesses' fees. Attorneys' fees may be awarded against the United States (or any agency or official of the United States) without regard to section 2412 of title 28, United States Code, or any other provision of law. For purposes of this paragraph and sections 11(c), 23(a), and 24, a reasonable attorney's fee is a fee (A) which is based upon (i) the actual time expended by an attorney in providing advice and other legal services in connection with representing a person in an action brought under this subsection, and (ii) such reasonable expenses as may be incurred by the attorney in the provision of such services, and (B) which is computed at the rate prevailing for the provision of similar services with respect to actions brought in the court which is awarding such fee.

(f) The remedies under this section shall be in addition to, and not in lieu of, other remedies provided by law.

(g) Subsection (e) of this section shall apply only with respect to petitions filed more than 3 years after the date of enactment of this Act.

JUDICIAL REVIEW OF CONSUMER PRODUCT SAFETY RULES

15 U.S.C. 2060

SEC. 11. (a) Not later than 60 days after a consumer product safety rule is promulgated by the Commission, any person adversely affected by such rule, or any consumer or consumer organization, may file a petition with the United States court of appeals for the District of Columbia or for the circuit in which such person, consumer, or organization resides or has his principal place of business for judicial review of such rule. Copies of the petition shall be forthwith transmitted by the clerk of the court to the Commission or other officer designated by it for that purpose and to the Attorney General. The record of the proceedings on which the Commission based its rule shall be filed in the court as provided for in section 2112 of title 28, United States Code.

For purposes of this section, the term "record" means such consumer product safety rule; any notice or proposal published pursuant to section 7, 8, or 9; the transcript required by section 9(a)(2) of any oral presentation; any written submission of interested parties; and any other information which the Commission considers relevant to such rule.

(b) If the petitioner applies to the court for leave to adduce additional data, views, or arguments and shows to the satisfaction of the court that such additional data, views, or arguments are material and that there were reasonable grounds for the petitioner's failure to adduce such data, views, or arguments in the proceeding before the Commission, the court may order the Commission to provide additional opportunity for the oral presentation of data, views, or arguments and for written submissions. The Commission may modify its findings, or make new findings by reason of the additional data, views, or arguments so taken and shall file such modified or new findings, and its recommendation, if any, for the modification or setting aside of its original rule, with the return of such additional data, views, or arguments.

(c) Upon the filing of the petition under subsection (a) of this section the court shall have jurisdiction to review the consumer product safety rule in accordance with chapter 7 of title 5, United States Code, and to grant appropriate relief, including interim relief, as provided in such chapter. A court may in the interest of justice include in such relief an award of the costs of suit, including reasonable attorneys' fees (determined in accordance with section 10(e)(4) and reasonable expert witnesses' fees. Attorneys' fees may be awarded against the United States (or any agency or official of the United States) without regard to section 2412 of title 28, United States Code, or any other provision of law. The consumer product safety rule shall not be affirmed unless the Commission's findings under section 9(c) are supported by substantial evidence on the record taken as a whole.

(d) The judgment of the court affirming or setting aside, in whole or in part, any consumer product safety rule shall be final, subject to review by the Supreme Court of the United States upon certiorari or certification, as provided in section 1254 of title 28 of the United States Code.

(e) The remedies provided for in this section shall be in addition to and not in lieu of any other remedies provided by law.

IMMINENT HAZARDS

SEC. 12. (a) The Commission may file in a United States district court an action (1) against an imminently hazardous consumer product for seizure of such product under subsection (b)(2), or (2) against any person who is a manufacturer, distributor, or retailer of such product, or (3) against both. Such an action may be filed notwithstanding the existence of a consumer product safety rule applicable to such product, or the pendency of any administrative or judicial proceedings under any other provision of this Act. As used in this section, and hereinafter in this Act, the term "imminently hazardous consumer product" means a consumer product which presents imminent and unreasonable risk of death, serious illness, or severe personal injury.

15 U.S.C. 2061

(b)(1) The district court in which such action is filed shall have jurisdiction to declare such product an imminently hazardous consumer product, and (in the case of an action under subsection (a)(2)) to grant (as ancillary to such declaration or in lieu thereof) such temporary or permanent relief as may be necessary to protect the public from such risk. Such relief may include a mandatory order requiring the notification of such risk to purchasers of such product known to the defendant, public notice, the recall, the repair or the replacement of, or refund for, such product.

(2) In the case of an action under subsection (a)(1), the consumer product may be proceeded against by process of libel for the seizure and condemnation of such product in any United States district court within the jurisdiction of which such consumer product is found. Proceedings and cases instituted under the authority of the preceding sentence shall conform as nearly as possible to proceedings in rem in admiralty.

(c) Where appropriate, concurrently with the filing of such action or as soon thereafter as may be practicable, the Commission shall initiate a proceeding to promulgate a consumer product safety rule applicable to the consumer product with respect to which such action is filed.

(d)(1) Prior to commencing an action under subsection (a), the Commission may consult the Product Safety Advisory Council (established under section 28) with respect to its determination to commence such action, and request the Council's recommendations as to the type of temporary or permanent relief which may be necessary to protect the public.

(2) The Council shall submit its recommendations to the Commission within one week of such request.

(3) Subject to paragraph (2), the Council may conduct such hearing or offer such opportunity for the presentation of views as it may consider necessary or appropriate.

(e)(1) An action under subsection (a)(2) of this section may be brought in the United States district court for the District of Columbia or in any judicial district in which any of the defendants is found, is an inhabitant or transacts business; and process in such an action may be served on a defendant in any other district in which such defendant resides or may be found. Subpenas requiring attendance of witnesses in such an action may run into any other district. In determining the judicial district in which an action may be brought under this section in instances in which such action may be brought in more than one judicial district, the Commission shall take into account the convenience of the parties.

(2) Whenever procedings under this section involving substantially similar consumer products are pending in courts in two or more judicial districts, they shall be consolidated for trial by order of any such court upon application reasonably made by any party in interest, upon notice to all other parties in interest.

(f) Notwithstanding any other provision of law, in any action under this section, the Commission may direct attorneys employed by it to appear and represent it.

NEW PRODUCTS

15 U.S.C. 2062

SEC. 13. (a) The Commission may, by rule, prescribe procedures for the purpose of insuring that the manufacturer of any new consumer product furnish notice and a description of such product to the Commission before its distribution in commerce.

(b) For purposes of this section, the term "new consumer product" means a consumer product which incorporates a design, material, or form of energy exchange which (1) has not previously been used substantially in consumer products and (2) as to which there exists a lack of information adequate to determine the safety of such product in use by consumers.

PRODUCT CERTIFICATION AND LABELING

SEC. 14. (a)(1) Every manufacturer of a product which is subject to a consumer product safety standard under this Act and which is distributed in commerce (and the private labeler of such product if it bears a private label) shall issue a certificate which shall certify that such product conforms to all applicable consumer product safety standards, and shall specify any standard which is applicable. Such certificate shall accompany the product or shall otherwise be furnished to any distributor or retailer to whom the product is delivered. Any certificate under this subsection shall be based on a test of each product or upon a reasonable testing program; shall state the name of the manufacturer or private labeler issuing the certificate; and shall include the date and place of manufacture.

15 U.S.C. 2063

(2) In the case of a consumer product for which there is more than one manufacturer or more than one private labeler, the Commission may by rule designate one or more of such manufacturers or one or more of such private labelers (as the case may be) as the persons who shall issue the certificate required by paragraph (1) of this subsection, and may exempt all other manufacturers of such product or all other private labelers of the product (as the case may be) from the requirement under paragraph (1) to issue a certificate with respect to such product.

(b) The Commission may by rule prescribe reasonable testing programs for consumer products which are subject to consumer product safety standards under this Act and for which a certificate is required under subsection (a). Any test or testing program on the basis of which a certificate is issued under subsection (a) may, at the option of the person required to certify the product, be conducted by an independent third party qualified to perform such tests or testing programs.

(c) The Commission may by rule require the use and prescribe the form and content of labels which contain the following information (or that portion of it specified in the rule)—

(1) The date and place of manufacture of any consumer product.

(2) A suitable identification of the manufacturer of the consumer product, unless the product bears a private label in which case it shall identify the private labeler and shall also contain a code mark which will permit the seller of such product to identify the manufacturer thereof to the purchaser upon his request.

(3) In the case of a consumer product subject to a consumer product safety rule, a certification that

the product meets all applicable consumer product safety standards and a specification of the standards which are applicable.

Such labels, where practicable, may be required by the Commission to be permanently marked on or affixed to any such consumer product. The Commission may, in appropriate cases, permit information required under paragraphs (1) and (2) of this subsection to be coded.

NOTIFICATION AND REPAIR, REPLACEMENT, OR REFUND

15 U.S.C. 2064

SEC. 15. (a) For purposes of this section, the term "substantial product hazard" means—

(1) a failure to comply with an applicable consumer product safety rule which creates a substantial risk of injury to the public, or

(2) a product defect which (because of the pattern of defect, the number of defective products distributed in commerce, the severity of the risk, or otherwise) creates a substantial risk of injury to the public.

(b) Every manufacturer of a consumer product distributed in commerce, and every distributor and retailer of such product, who obtains information which reasonably supports the conclusion that such product—

(1) fails to comply with an applicable consumer product safety rule; or

(2) contains a defect which could create a substantial product hazard described in subsection (a) (2),

shall immediately inform the Commission of such failure to comply or of such defect, unless such manufacturer, distributor, or retailer has actual knowlege that the Commission has been adequately informed of such defect or failure to comply.

(c) If the Commission determines (after affording interested persons, including consumers and consumer organizations, an opportunity for a hearing in accordance with subsection (f) of this section) that a product distributed in commerce presents a substantial product hazard and that notification is required in order to adequately protect the public from such substantial product hazard, the Commission may order the manufacturer or any distributor or retailer of the product to take any one or more of the following actions:

(1) To give public notice of the defect or failure to comply.

(2) To mail notice to each person who is a manufacturer, distributor, or retailer of such product.

(3) To mail notice to every person to whom the person required to give notice knows such product was delivered or sold.

Any such order shall specify the form and content of any notice required to be given under such order.

(d) If the Commission determines (after affording interested parties, including consumers and consumer organizations, an opportunity for a hearing in accordance with subsection (f)) that a product distributed in commerce presents a substantial product hazard and that action under this subsection is in the public interest, it may order the manufacturer or any distributor or retailer of such product to take whichever of the following actions the person to whom the order is directed elects:

(1) To bring such product into conformity with the requirements of the applicable consumer product safety rule or to repair the defect in such product.

(2) To replace such product with a like or equivalent product which complies with the applicable consumer product safety rule or which does not contain the defect.

(3) To refund the purchase price of such product (less a reasonable allowance for use, if such product has been in the possession of a consumer for one year or more (A) at the time of public notice under subsection (c), or (B) at the time the consumer receives actual notice of the defect or noncompliance, whichever first occurs).

An order under this subsection may also require the person to whom it applies to submit a plan, satisfactory to the Commission, for taking action under whichever of the preceding paragraphs of this subsection under which such person has elected to act. The Commission shall specify in the order the persons to whom refunds must be made if the person to whom the order is directed elects to take the action described in paragraph (3). If an order under this subsection is directed to more than one person, the Commission shall specify which person has the election under this subsection. An order under this subsection may prohibit the person to whom it applies from manufacturing for sale, offering for sale, distributing in commerce, or importing into the customs territory of the United States (as defined in general headnote 2 to the Tariff Schedules of the United States), or from doing any combination of such actions, the product with respect to which the order was issued.

(e)(1) No charge shall be made to any person (other than a manufacturer, distributor, or retailer) who avails himself of any remedy provided under an order issued under subsection (d), and the person subject to the order shall reimburse each person (other than a manufacturer, distributor, or retailer) who is entitled to such a remedy for any reasonable and foreseeable expenses incurred by such person in availing himself of such remedy.

(2) An order issued under subsection (c) or (d) with respect to a product may require any person who is a manufacturer, distributor, or retailer of the product to reimburse any other person who is a manufacturer, distributor, or retailer of such product for such other person's expenses in connection with carrying out the order, if the Commission determines such reimbursement to be in the public interest.

(f) An order under subsection (c) or (d) may be issued only after an opportunity for a hearing in accordance with section 554 of title 5, United States Code, except that, if the Commission determines that any person who wishes to participate in such hearing is a part of a class of participants who share an identity of interest, the Commission may limit such person's participation in such hearing to participation through a single representative designated by such class (or by the Commission if such class fails to designate such a representative).

(g)(1) If the Commission has initiated a proceeding under this section for the issuance of an order under subsection (d) with respect to a product which the Commission has reason to believe presents a substantial product hazard, the Commission (without regard to section 27(b)(7)), or the Attorney General may, in accordance with section 12(e)(1), apply to a district court of the United States for the issuance of a preliminary injunction to restrain the distribution in commerce of such product pending the completion of such proceeding. If such a preliminary injunction has been issued, the Commission (or the Attorney General if the preliminary injunction was issued upon an application of the Attorney General) may apply to the issuing court for extensions of such preliminary injunction.

(2) Any preliminary injunction, and any extension of a preliminary injunction, issued under this subsection with respect to a product shall be in effect for such period as the issuing court prescribes not to exceed a period which extends beyond the thirtieth day from the date of the issuance of the preliminary injunction (or, in the case of a preliminary injunction which has been extended, the date of its extension) or the date of the completion or termination of the proceeding under this section respecting such product, whichever date occurs first.

(3) The amount in controversy requirement of section 1331 of title 28, United States Code, does not apply with respect to the jurisdiction of a district court of the United States to issue or extend a preliminary injunction under this subsection.

INSPECTION AND RECORDKEEPING

SEC. 16. (a) For purposes of implementing this Act, or rules or orders prescribed under this Act, officers or employees duly designated by the Commission, upon presenting appropriate credentials and a written notice from the Commission to the owner, operator, or agent in charge, are authorized— 15 U.S.C. 2065

(1) to enter, at reasonable times, (A) any factory, warehouse, or establishment in which consumer products are manufactured or held, in connection with distribution in commerce, or (B) any conveyance being used to transport consumer products in connection with distribution in commerce; and

(2) to inspect, at reasonable times and in a reasonable manner such conveyance or those areas of such factory, warehouse, or establishment where such products are manufactured, held, or transported and which may relate to the safety of such products. Each such inspection shall be commenced and completed with reasonable promptness.

(b) Every person who is a manufacturer, private labeler, or distributor of a consumer product shall establish and maintain such records, make such reports, and provide such information as the Commission may, by rule, reasonably require for the purposes of implementing this Act, or to determine compliance with rules or orders prescribed under this Act. Upon request of an officer or employee duly designated by the Commission, every such manufacturer, private labeler, or distributor shall permit the inspection of appropriate books, records, and papers relevant to determining whether such manufacturer, private labeler, or distributor has acted or is acting in compliance with this Act and rules under this Act.

IMPORTED PRODUCTS

SEC. 17. (a) Any consumer product offered for importation into the customs territory of the United States (as defined in general headnote 2 to the Tariff Schedules of the United States) shall be refused admission into such customs territory if such product— 15 U.S.C. 2066

(1) fails to comply with an applicable consumer product safety rule;

(2) is not accompanied by a certificate required by section 14, or is not labeled in accordance with regulations under section 14(c);

(3) is or has been determined to be an imminently hazardous consumer product in a proceeding brought under section 12;

(4) has a product defect which constitutes a substantial product hazard (within the meaning of section 15(a)(2)); or

(5) is a product which was manufactured by a person who the Commission has informed the Secretary of the Treasury is in violation of subsection (g).

(b) The Secretary of the Treasury shall obtain without charge and deliver to the Commission, upon the latter's request, a reasonable number of samples of consumer products being offered for import. Except for those owners or consignees who are or have been afforded an opportunity for a hearing in a proceeding under section 12 with respect to an imminently hazardous product, the owner or consignee of the product shall be afforded an opportunity by the Commission for a hearing in accordance with section 554 of title 5 of the United States Code with respect to the importation of such products into the customs territory of the United States. If it appears from examination of such samples or otherwise that a product must be refused admission under the terms of subsection (a), such product shall be refused admission, unless subsection (c) of this section applies and is complied with.

(c) If it appears to the Commission that any consumer product which may be refused admission pursuant to subsection (a) of this section can be so modified that it need not (under the terms of paragraphs (1) through (4) of subsection (a)) be refused admission, the Commission may defer final determination as to the admission of such product and, in accordance with such regulations as the Commission and the Secretary of the Treasury shall jointly agree to, permit such product to be delivered from customs custody under bond for the purpose of permitting the owner or consignee an opportunity to so modify such product.

(d) All actions taken by an owner or consignee to modify such product under subsection (c) shall be subject to the supervision of an officer or employee of the Commission and of the Department of the Treasury. If it appears to the Commission that the product cannot be so modified or that the owner or consignee is not proceeding satisfactorily to modify such product, it shall be refused admission into the customs territory of the United States, and the Commission may direct the Secretary to demand redelivery of the product into customs custody, and to seize the product in accordance with section 22(b) if it is not so redelivered.

(e) Products refused admission into the customs territory of the United States under this section must be exported, except that upon application, the Secretary of the Treasury may permit the destruction of the product in lieu of exportation. If the owner or consignee does not export the product within a reasonable time, the Department of the Treasury may destroy the product.

(f) All expenses (including travel, per diem or subsistence, and salaries of officers or employees of the United States) in connection with the destruction provided for in this section (the amount of such expenses to be determined in accordance with regulations of the Secretary of the Treasury) and all expenses in connection with the storage, cartage, or labor with respect to any consumer product refused admission under this section, shall be paid by the owner or consignee and, in default of such payment, shall constitute a lien against any future importations made by such owner or consignee.

(g) The Commission may, by rule, condition the importation of a consumer product on the manufacturer's compliance with the inspection and recordkeeping requirements of this Act and the Commission's rules with respect to such requirements.

EXPORTS

SEC. 18. This Act shall not apply to any consumer product if (1) it can be shown that such product is manufactured, sold, or held for sale for export from the United States (or that such product was imported for export), unless such consumer product is in fact distributed in commerce for use in the United States, and (2) such consumer product when distributed in commerce, or any container in which it is enclosed when so distributed, bears a stamp or label stating that such consumer product is intended for export; except that this Act shall apply to any consumer product manufactured for sale, offered for sale, or sold for shipment to any installation of the United States located outside of the United States.

15 U.S.C. 2067

PROHIBITED ACTS

SEC. 19. (a) It shall be unlawful for any person to—

15 U.S.C. 2068

(1) manufacture for sale, offer for sale, distribute in commerce, or import into the United States any consumer product which is not in conformity with an applicable consumer product safety standard under this Act;

(2) manufacture for sale, offer for sale, distribute in commerce, or import into the United States any consumer product which has been declared a banned hazardous product by a rule under this Act;

(3) fail or refuse to permit access to or copying of records, or fail or refuse to establish or maintain records, or fail or refuse to make reports or provide information, or fail or refuse to permit entry or

inspection, as required under this Act or rule thereunder;

(4) fail to furnish information required by section 15(b);

(5) fail to comply with an order issued under section 15 (c) or (d) (relating to notification, and to repair, replacement, and refund, and to prohibited acts);

(6) fail to furnish a certificate required by section 14 or issue a false certificate if such person in the exercise of due care has reason to know that such certificate is false or misleading in any material respect; or to fail to comply with any rule under section 14(c) (relating to labeling);

(7) fail to comply with any rule under section 9(d)(2) (relating to stockpiling);

(8) fail to comply with any rule under section 13 (relating to prior notice and description of new consumer products); or

(9) fail to comply with any rule under section 27(e) (relating to provision of performance and technical data).

(b) Paragraphs (1) and (2) of subsection (a) of this section shall not apply to any person (1) who holds a certificate issued in accordance with section 14(a) to the effect that such consumer product conforms to all applicable consumer product safety rules, unless such person knows that such consumer product does not conform, or (2) who relies in good faith on the representation of the manufacturer or a distributor of such product that the product is not subject to an applicable product safety rule.

CIVIL PENALTIES

15 U.S.C. 2069

SEC. 20. (a)(1) Any person who knowingly violates section 19 of this Act shall be subject to a civil penalty not to exceed $2,000 for each such violation. Subject to paragraph (2), a violation of section 19(a) (1), (2), (4), (5), (6), (7), (8), or (9) shall constitute a separate offense with respect to each consumer product involved, except that the maximum civil penalty shall not exceed $500,000 for any related series of violations. A violation of section 19(a)(3) shall constitute a separate violation with respect to each failure or refusal to allow or perform an act required thereby; and, if such violation is a continuing one, each day of such violation shall constitute a separate offense, except that the maximum civil penalty shall not exceed $500,000 for any related series of violations.

(2) The second sentence of paragraph (1) of this subsection shall not apply to violations of paragraph (1) or (2) of section 19(a)—

(A) if the person who violated such paragraphs is not the manufacturer or private labeler or a distributor of the products involved, and

(B) if such person did not have either (i) actual knowledge that his distribution or sale of the product violated such paragraphs or (ii) notice from the Commission that such distribution or sale would be a violation of such paragraphs.

(b) Any civil penalty under this section may be compromised by the Commission. In determining the amount of such penalty or whether it should be remitted or mitigated and in what amount, the appropriateness of such penalty to the size of the business of the person charged and the gravity of the violation shall be considered. The amount of such penalty when finally determined, or the amount agreed on compromise, may be deducted from any sums owing by the United States to the person charged.

(c) As used in the first sentence of subsection (a)(1) of this section, the term "knowingly" means (1) the having of actual knowledge, or (2) the presumed having of knowledge deemed to be possessed by a reasonable man who acts in the circumstances, including knowledge obtainable upon the exercise of due care to ascertain the truth of representations.

CRIMINAL PENALTIES

SEC. 21. (a) Any person who knowingly and willfully violates section 19 of this Act after having received notice of noncompliance from the Commission shall be fined not more than $50,000 or be imprisoned not more than one year, or both. 15 U.S.C. 2070

(b) Any individual director, officer, or agent of a corporation who knowingly and willfully authorizes, orders, or performs any of the acts or practices constituting in whole or in part a violation of section 19, and who has knowledge of notice of noncompliance received by the corporation from the Commission, shall be subject to penalties under this section without regard to any penalties to which that corporation may be subject under subsection (a).

INJUNCTIVE ENFORCEMENT AND SEIZURE

SEC. 22. (a) The United States district courts shall have jurisdiction to take the following action: 15 U.S.C. 2071

(1) Restrain any violation of section 19.

(2) Restrain any person from manufacturing for sale, offering for sale, distributing in commerce, or importing into the United States a product in violation of an order in effect under section 15(d).

(3) Restrain any person from distributing in commerce a product which does not comply with a consumer product safety rule. Such actions may be brought by the Commission (without regard to section 27(b)(7)(A)) or by the Attorney General in any United States district court for a district wherein any act, omission, or transaction constituting the violation occurred, or in such court for the district wherein the defendant is found or transacts business. In any action under this section process may be served on a defendant in any other district in which the defendant resides or may be found.

(b) Any consumer product—

(1) which fails to conform with an applicable consumer product safety rule, or

(2) the manufacture for sale, offering for sale, distribution in commerce, or the importation into the United States of which has been prohibited by an order in effect under section 15(d),

when introduced into or while in commerce or while held for sale after shipment in commerce shall be liable to be proceeded against on libel of information and condemned in any district court of the United States within the jurisdiction of which such consumer product is found. Proceedings in cases instituted under the authority of this subsection shall conform as nearly as possible to proceedings in rem in admiralty. Whenever such proceedings involving substantially similar consumer products are pending in courts of two or more judicial districts they shall be consolidated for trial by order of any such court upon application reasonably made by any party in interest upon notice to all other parties in interest.

SUITS FOR DAMAGES BY PERSONS INJURED

15 U.S.C. 2072

Sec. 23. (a) Any person who shall sustain injury by reason of any knowing (including willful) violation of a consumer product safety rule, or any other rule or order issued by the Commission may sue any person who knowingly (including willfully) violated any such rule or order in any district court of the United States in the district in which the defendant resides or is found or has an agent, subject to the provisions of section 1331 of title 28, United States Code as to the amount in controversy, shall recover damages sustained, and may, if the court determines it to be in the interest of justice, recover the costs of suit, including reasonable attorneys' fees (determined in accordance with section 10(e)(4)) and reasonable expert witnesses' fees.

(b) The remedies provided for in this section shall be in addition to and not in lieu of any other remedies provided by common law or under Federal or State law.

PRIVATE ENFORCEMENT OF PRODUCT SAFETY RULES AND OF SECTION 15 ORDERS

SEC. 24. Any interested person may bring an action in any United States district court for the district in which the defendant is found or transacts business to enforce a consumer product safety rule or an order under section 15, and to obtain appropriate injunctive relief. Not less than thirty days prior to the commencement of such action, such interested person shall give notice by registered mail to the Commission, to the Attorney General, and to the person against whom such action is directed. Such notice shall state the nature of the alleged violation of any such standard or order, the relief to be requested, and the court in which the action will be brought. No separate suit shall be brought under this section if at the time the suit is brought the same alleged violation is the subject of a pending civil or criminal action by the United States under this Act. In any action under this section the court may in the interest of justice award the costs of suit, including reasonable attorneys' fees (determined in accordance with section 10(e)(4)) and reasonable expert witnesses' fees.

15 U.S.C. 2073

EFFECT ON PRIVATE REMEDIES

SEC. 25. (a) Compliance with consumer product safety rules or other rules or orders under this Act shall not relieve any person from liability at common law or under State statutory law to any other person.

15 U.S.C. 2074

(b) The failure of the Commission to take any action or commence a proceeding with respect to the safety of a consumer product shall not be admissible in evidence in litigation at common law or under State statutory law relating to such consumer product.

(c) Subject to sections 6(a)(2) and 6(b) but notwithstanding section 6(a)(1), (1) any accident or investigation report made under this Act by an officer or employee of the Commission shall be made available to the public in a manner which will not identify any injured person or any person treating him, without the consent of the person so identified, and (2) all reports on research projects, demonstration projects, and other related activities shall be public information.

EFFECT ON STATE STANDARDS

SEC. 26. (a) Whenever a consumer product safety standard under this Act is in effect and applies to a risk of injury associated with a consumer product, no State or political subdivision of a State shall have any authority either to establish or to continue in effect any

15 U.S.C. 2075

provision of a safety standard or regulation which prescribes any requirements as to the performance, composition, contents, design, finish, construction, packaging, or labeling of such product which are designed to deal with the same risk of injury associated with such consumer product, unless such requirements are identical to the requirements of the Federal standard.

(b) Subsection (a) of this section does not prevent the Federal Government or the government of any State or political subdivision of a State from establishing or continuing in effect a safety requirement applicable to a consumer product for its own use which requirement is designed to protect against a risk of injury associated with the product and which is not identical to the consumer product safety standard applicable to the product under this Act if the Federal, State, or political subdivision requirement provides a higher degree of protection from such risk of injury than the standard applicable under this Act.

(c) Upon application of a State or political subdivision of a State, the Commission may by rule, after notice and opportunity for oral presentation of views, exempt from the provisions of subsection (a) (under such conditions as it may impose in the rule) any proposed safety standard or regulation which is described in such application and which is designed to protect against a risk of injury associated with a consumer product subject to a consumer product safety standard under this Act if the State or political subdivision standard or regulation—

(1) provides a significantly higher degree of protection from such risk of injury than the consumer product safety standard under this Act, and

(2) does not unduly burden interstate commerce.

In determining the burden, if any, of a State or political subdivision standard or regulation on interstate commerce, the Commission shall consider and make appropriate (as determined by the Commission in its discretion) findings on the technological and economic feasibility of complying with such standard or regulation, the cost of complying with such standard or regulation, the geographic distribution of the consumer product to which the standard or regulation would apply, the probability of other States or political subdivisions applying for an exemption under this subsection for a similar standard or regulation, and the need for a national, uniform standard under this Act for such consumer product.

ADDITIONAL FUNCTIONS OF COMMISSION

15 U.S.C. 2076

SEC. 27. (a) The Commission may, by one or more of its members or by such agents or agency as it may designate, conduct any hearing or other inquiry necessary or appropriate to its functions anywhere in the United States. A Commissioner who participates in such a hearing or other inquiry shall not be disqualified solely by reason of such participation from subsequently participating in a decision of the Commission in the same matter. The Commission shall publish notice of any proposed hearing in the Federal Register and shall afford a reasonable opportunity for interested persons to present relevant testimony and data.

(b) The Commission shall also have the power—

(1) to require, by special or general orders, any person to submit in writing such reports and answers to questions as the Commission may prescribe; and such submission shall be made within such reasonable period and under oath or otherwise as the Commission may determine;

(2) to administer oaths;

(3) to require by subpena the attendance and testimony of witnesses and the production of all documentary evidence relating to the execution of its duties;

(4) in any proceeding or investigation to order testimony to be taken by deposition before any person who is designated by the Commission and has the power to administer oaths and, in such instances, to compel testimony and the production of evidence in the same manner as authorized under paragraph (3) of this subsection;

(5) to pay witnesses the same fees and mileage as are paid in like circumstances in the courts of the United States;

(6) to accept gifts and voluntary and uncompensated services, notwithstanding the provisions of section 3679 of the Revised Statutes (31 U.S.C. 665 (b));

(7) to —

(A) initiate, prosecute, defend, or appeal (other than to the Supreme Court of the United States), through its own legal representative and in the name of the Commission, any civil action if the Commission makes a written request to the Attorney General for representation in such civil action and the Attorney General does not within the 45-day period beginning on the date such request was made notify the Commission in writing that the Attorney General will represent the Commission in such civil action, and

(B) initiate, prosecute, or appeal, through its own legal representative, with the concurrence of the Attorney General or through the Attorney General, any criminal action, for the purpose of enforcing the laws subject to its jurisdiction;

(8) to lease buildings or parts of buildings in the District of Columbia, without regard to the Act of March 3, 1877 (40 U.S.C. 34), for the use of the Commission; and

(9) to delegate any of its functions or powers, other than the power to issue subpenas under paragraph (3), to any officer or employee of the Commission.

(c) Any United States district court within the jurisdiction of which any inquiry is carried on, may, upon petition by the Commission (subject to subsection (b)(7)) or by the Attorney General, in case of refusal to obey a subpena or order of the Commission issued under subsection (b) of this section, issue an order requiring compliance therewith; and any failure to obey the order of the court may be punished by the court as a contempt thereof.

(d) No person shall be subject to civil liability to any person (other than the Commission or the United States) for disclosing information at the request of the Commission.

(e) The Commission may by rule require any manufacturer of consumer products to provide to the Commission such performance and technical data related to performance and safety as may be required to carry out the purposes of this Act, and to give such notification of such performance and technical data at the time of original purchase to prospective purchasers and to the first purchaser of such product for purposes other than resale, as it determines necessary to carry out the purposes of this Act.

(f) For purposes of carrying out this Act, the Commission may purchase any consumer product and it may require any manufacturer, distributor, or retailer of a consumer product to sell the product to the Commission at manufacturer's, distributor's, or retailer's cost.

(g) The Commission is authorized to enter into contracts with governmental entities, private organizations, or individuals for the conduct of activities authorized by this Act.

(h) The Commission may plan, construct, and operate a facility or facilities suitable for research, development, and testing of consumer products in order to carry out this Act.

(i)(1) Each recipient of assistance under this Act pursuant to grants or contracts entered into under other than competitive bidding procedures shall keep such records as the Commission by rule shall prescribe, including records which fully disclose the amount and disposition by such recipient of the proceeds of such assistance, the total cost of the project undertaken in connection with which such assistance is given or used, and the amount of that portion of the cost of the project or undertaking supplied by other sources, and such other records as will facilitate an effective audit.

(2) The Commission and the Comptroller General of the United States, or their duly authorized representatives, shall have access for the purpose of audit and examination to any books, documents, papers, and records of the recipients that are pertinent to the grants or contracts entered into under this Act under other than competitive bidding procedures.

(j) The Commission shall prepare and submit to the President and the Congress at the beginning of each regular session of Congress a comprehensive report on the administration of this Act for the preceding fiscal year. Such report shall include—

(1) a thorough appraisal, including statistical analyses, estimates, and long-term projections, of the incidence of injury and effects to the population resulting from consumer products, with a breakdown, insofar as practicable, among the various sources of such injury;

(2) a list of consumer product safety rules prescribed or in effect during such year;

(3) an evaluation of the degree of observance of consumer product safety rules, including a list of enforcement actions, court decisions, and compromises of alleged violations, by location and company name;

(4) a summary of outstanding problems confronting the administration of this Act in order of priority;

(5) an analysis and evaluation of public and private consumer product safety research activities;

(6) a list, with a brief statement of the issues, of completed or pending judicial actions under this Act;

(7) the extent to which technical information was disseminated to the scientific and commercial communities and consumer information was made available to the public;

(8) the extent of cooperation between Commission officials and representatives of industry and other interested parties in the implementation of this Act, including a log or summary of meetings held between Commission officials and representatives of industry and other interested parties;

(9) an appraisal of significant actions of State and local governments relating to the responsibilities of the Commission; and

(10) such recommendations for additional legislation as the Commission deems necessary to carry out the purposes of this Act.

(k) (1) Whenever the Commission submits any budget estimate or request to the President or the Office of Management and Budget, it shall concurrently transmit a copy of that estimate or request to the Congress.

(2) Whenever the Commission submits any legislative recommendations, or testimony, or comments on legislation to the President or the Office of Management and Budget, it shall concurrently transmit a copy thereof to the Congress. No officer or agency of the United States shall have any authority to require the Commission to submit its legislative recommendations, or testimony, or comments on legislation, to any officer or agency of the United States for approval, comments, or review, prior to the submission of such recommendations, testimony, or comments to the Congress.

(l) (1) Except as provided in paragraph (2)—

(A) the Commission shall transmit to the Committee on Commerce of the Senate and the Committee on Interstate and Foreign Commerce of the House of Representatives each consumer product safety rule proposed after the date of the enactment of this subsection and each regulation proposed by the Commission after such date under section 2 or 3 of the Federal Hazardous Substances Act, section 3 of the Poison Prevention Packaging Act of 1970, or section 4 of the Flammable Fabrics Act; and

(B) no consumer product safety rule and no regulation under a section referred to in subparagraph (A) may be adopted by the Commission before the thirtieth day after the date the proposed rule or regulation upon which such rule or regulation was based was transmitted pursuant to subparagraph (A).

(2) Paragraph (1) does not apply with respect to a regulation under section 2(q) of the Federal Hazardous Substances Act respecting a hazardous substance the distribution of which is found under paragraph (2) of such section to present an imminent hazard or a regulation under section 3(e) of such Act respecting a toy or other article intended for use by children the distribution of which is found under paragraph (2) of such section to present an imminent hazard.

PRODUCT SAFETY ADVISORY COUNCIL

15 U.S.C. 2077

SEC. 28. (a) The Commission shall establish a Product Safety Advisory Council which it may consult before prescribing a consumer product safety rule or taking other action under this Act. The Council shall be appointed by the Commission and shall be composed of fifteen members, each of whom shall be qualified by training and experience in one or more of the fields applicable to the safety of products within the jurisdiction of the Commission. The Council shall be constituted as follows:

(1) five members shall be selected from governmental agencies including Federal, State, and local governments;

(2) five members shall be selected from consumer product industries including at least one representative of small business; and

(3) five members shall be selected from among consumer organizations, community organizations, and recognized consumer leaders.

(b) The Council shall meet at the call of the Commission, but not less often than four times during each calendar year.

(c) The Council may propose consumer product safety rules to the Commission for its consideration and may function through subcommittees of its members. All proceedings of the Council shall be public, and a record of each proceeding shall be available for public inspection.

(d) Members of the Council who are not officers or employees of the United States shall, while attending meetings or conferences of the Council or while otherwise engaged in the business of the Council, be entitled to receive compensation at a rate fixed by the Commission, not exceeding the daily equivalent of the annual rate of basic pay in effect for grade GS–18 of the General Schedule, including traveltime, and while away from their homes or regular places of business they may be allowed travel expenses, including per diem in lieu of subsistence, as authorized by section 5703 of title 5, United States Code. Payments under this subsection shall not render members of the Council officers or employees of the United States for any purpose.

COOPERATION WITH STATES AND WITH OTHER FEDERAL AGENCIES

SEC. 29. (a) The Commission shall establish a program to promote Federal-State cooperation for the purposes of carrying out this Act. In implementing such program the Commission may—

15 U.S.C. 2078

(1) accept from any State or local authorities engaged in activities relating to health, safety, or consumer protection assistance in such functions as injury data collection, investigation, and educational programs, as well as other assistance in the administration and enforcement of this Act which such States or localities may be able and willing to pro-

vide and, if so agreed, may pay in advance or otherwise for the reasonable cost of such assistance, and

(2) commission any qualified officer or employee of any State or local agency as an officer of the Commission for the purpose of conducting examinations, investigations, and inspections.

(b) In determining whether such proposed State and local programs are appropriate in implementing the purposes of this Act, the Commission shall give favorable consideration to programs which establish separate State and local agencies to consolidate functions relating to product safety and other consumer protection activities.

(c) The Commission may obtain from any Federal department or agency such statistics, data, program reports, and other materials as it may deem necessary to carry out its functions under this Act. Each such department or agency may cooperate with the Commission and, to the extent permitted by law, furnish such materials to it. The Commission and the heads of other departments and agencies engaged in administering programs related to product safety shall, to the maximum extent practicable, cooperate and consult in order to insure fully coordinated efforts.

(d) The Commission shall, to the maximum extent practicable, utilize the resources and facilities of the National Bureau of Standards, on a reimbursable basis, to perform research and analyses related to risks of injury associated with consumer products (including fire and flammability risks), to develop test methods, to conduct studies and investigations, and to provide technical advice and assistance in connection with the functions of the Commission.

(e) The Commission may provide to another Federal agency or a State or local agency or authority engaged in activities relating to health, safety, or consumer protection, copies of any accident or investigation report made under this Act by any officer, employee, or agent of the Commission only if (1) information which under section 6(a)(2) is to be considered confidential is not included in any copy of such report which is provided under this subsection; and (2) each Federal agency and State and local agency and authority which is to receive under this subsection a copy of such report provides assurances satisfactory to the Commission that the identity of any injured person and any person who treated an injured person will not, without the consent of the person identified, be included in—

(A) any copy of any such report, or

(B) any information contained in any such report,

which the agency or authority makes available to any member of the public. No Federal agency or State or local agency or authority may disclose to the public any information contained in a report received by the agency or authority under this subsection unless with respect to such information the Commission has complied with the applicable requirements of section 6(b).

TRANSFERS OF FUNCTIONS

SEC. 30. (a) The functions of the Secretary of Health, Education, and Welfare under the Federal Hazardous Substances Act (15 U.S.C. 1261 et seq.) and the Poison Prevention Packaging Act of 1970 are transferred to the Commission. The functions of the Secretary of Health, Education, and Welfare under the Federal Food, Drug, and Cosmetic Act (15 U.S.C. 301 et seq.), to the extent such functions relate to the administration and enforcement of the Poison Prevention Packaging Act of 1970, are transferred to the Commission.

15 U.S.C. 2079

(b) The functions of the Secretary of Health, Education, and Welfare, the Secretary of Commerce, and the Federal Trade Commission under the Flammable Fabrics Act (15 U.S.C. 1191 et seq.) are transferred to the Commission. The functions of the Federal Trade Commission under the Federal Trade Commission Act, to the extent such functions relate to the administration and enforcement of the Flammable Fabrics Act, are transferred to the Commission.

(c) The functions of the Secretary of Commerce and the Federal Trade Commission under the Act of August 2, 1956 (15 U.S.C. 1211) are transferred to the Commission.

(d) A risk of injury which is associated with a consumer product and which could be eliminated or reduced to a sufficient extent by action under the Federal Hazardous Substances Act, the Poison Prevention Packaging Act of 1970, or the Flammable Fabrics Act may be regulated under this Act only if the Commission by rule finds that it is in the public interest to regulate such risk of injury under this Act. Such a rule shall identify the risk of injury proposed to be regulated under this Act and shall be promulgated in accordance with section 553 of title 5, United States Code; except that the period to be provided by the Commission pursuant to subsection (c) of such section for the submission of data, views, and arguments respecting the rule shall not exceed thirty days from the date of publication pursuant to subsection (b) of such section of a notice respecting the rule.

(e)(1)(A) All personnel, property, records, obligations, and commitments which are used primarily with respect to any function transferred under the provisions of subsections (a), (b) and (c) of this section shall be transferred to the Commission, except those associated with fire and flammability research in the National Bureau of Standards. The transfer of personnel pursuant to this paragraph shall be without reduction in classification or compensation for one year after such transfer, except that the Chairman of the Commission shall have full authority to assign personnel during such one-year period in order to efficiently carry out functions transferred to the Commission under this section.

(B) Any commissioned officer of the Public Health Service who upon the day before the effective date of this section, is serving as such officer primarily in the performance of functions transferred by this Act to the Commission, may, if such officer so elects, acquire competitive status and be transferred to a competitive position in the Commission subject to subparagraph (A) of this paragraph, under the terms prescribed in paragraphs (3) through (8)(A) of section 15(b) of the Clean Air Amendments of 1970 (84 Stat. 1676; 42 U.S.C. 215 nt).

(2) All orders, determinations, rules, regulations, permits, contracts, certificates, licenses, and privileges (A) which have been issued, made, granted, or allowed to become effective in the exercise of functions which are transferred under this section by any department or agency, any functions of which are transferred by this section, and (B) which are in effect at the time this section takes effect, shall continue in effect according to their terms until modified, terminated, superseded, set aside, or repealed by the Commission, by any court of competent jurisdiction, or by operation of law.

(3) The provisions of this section shall not affect any proceedings pending at the time this section takes effect before any department or agency, functions of which are transferred by this section; except that such proceedings, to the extent that they relate to functions so transferred, shall be continued before the Commission. Orders shall be issued in such proceedings, appeals shall be taken therefrom, and payments shall be made pursuant to such orders, as if this section had not been enacted; and orders issued in any such proceedings shall continue in effect until modified, terminated, superseded, or repealed by the Commission, by a court of competent jurisdiction, or by operation of law.

(4) The provisions of this section shall not affect suits commenced prior to the date this section takes effect and in all such suits proceedings shall be had, appeals taken, and judgments rendered, in the same manner and effect as if this section had not been enacted; except that if before the date on which this section takes effect, any department or agency (or officer thereof in his official capacity) is a party to a suit involving functions transferred to the Commission, then such suit shall be continued by the Commission. No cause of action, and no suit, action, or other proceeding, by or against any department or agency (or officer thereof in his official

capacity) functions of which are transferred by this section, shall abate by reason of the enactment of this section. Causes of actions, suits, actions, or other proceedings may be asserted by or against the United States or the Commission as may be appropriate and, in any litigation pending when this section takes effect, the court may at any time, on its own motion or that of any party, enter an order which will give effect to the provisions of this paragraph.

(f) For purposes of this section, (1) the term "function" includes power and duty, and (2) the transfer of a function, under any provision of law, of an agency or the head of a department shall also be a transfer of all functions under such law which are exercised by any office or officer of such agency or department.

LIMITATION ON JURISDICTION

SEC. 31. The Commission shall have no authority under this Act to regulate any risk of injury associated with a consumer product if such risk could be eliminated or reduced to a sufficient extent by actions taken under the Occupational Safety and Health Act of 1970; the Atomic Energy Act of 1954; or the Clean Air Act. The Commission shall have no authority under this Act to regulate any risk of injury associated with electronic product radiation emitted from an electronic product (as such terms are defined by sections 355 (1) and (2) of the Public Health Service Act) if such risk of injury may be subjected to regulation under subpart 3 of part F of title III of the Public Health Service Act.

15 U.S.C. 2080

AUTHORIZATION OF APPROPRIATIONS

SEC. 32. (a) There are authorized to be appropriated for the purposes of carrying out the provisions of this Act (other than the provisions of section 27(h) which authorize the planning and construction of research, development, and testing facilities) and for the purpose of carrying out the functions, powers, and duties transferred to the Commission under section 30, not to exceed—

15 U.S.C. 2081

(1) $51,000,000 for the fiscal year ending June 30, 1976;

(2) $14,000,000 for the period beginning July 1, 1976, and ending September 30, 1976;

(3) $60,000,000 for the fiscal year ending September 30, 1977; and

(4) $68,000,000 for the fiscal year ending September 30, 1978.

(b)(1) There are authorized to be appropriated such sums as may be necessary for the planning and construction of research, development and testing facilities de-

scribed in section 27(h); except that no appropriation shall be made for any such planning or construction involving an expenditure in excess of $100,000 if such planning or construction has not been approved by resolutions adopted in substantially the same form by the Committee on Interstate and Foreign Commerce of the House of Representatives, and by the Committee on Commerce of the Senate. For the purpose of securing consideration of such approval the Commission shall transmit to Congress a prospectus of the proposed facility including (but not limited to)—

(A) a brief description of the facility to be planned or constructed;

(B) the location of the facility, and an estimate of the maximum cost of the facility;

(C) a statement of those agencies, private and public, which will use such facility, together with the contribution to be made by each such agency toward the cost of such facility; and

(D) a statement of justification of the need for such facility.

(2) The estimated maximum cost of any facility approved under this subsection as set forth in the prospectus may be increased by the amount equal to the percentage increase, if any, as determined by the Commission, in construction costs, from the date of the transmittal of such prospectus to Congress, but in no event shall the increase authorized by this paragraph exceed 10 per centum of such estimated maximum cost.

(c) No funds appropriated under subsection (a) may be used to pay any claim described in section 4(i) whether pursuant to a judgment of a court or under any award, compromise, or settlement of such claim made under section 2672 of title 28, United States Code, or under any other provision of law.

SEPARABILITY

15 U.S.C. 2051 note

SEC. 33. If any provision of this Act, or the application of such provision to any person or circumstance, shall be held invalid, the remainder of this Act, or the application of such provisions to persons or circumstances other than those as to which it is held invalid, shall not be affected thereby.

EFFECTIVE DATE

15 U.S.C. 2051 note

SEC. 34. This Act shall take effect on the sixtieth day following the date of its enactment, except—

(1) sections 4 and 32 shall take effect on the date of enactment of this Act, and

(2) section 30 shall take effect on the later of (A) 150 days after the date of enactment of this Act, or (B) the date on which at least three memebrs of the Commission first take office.

Appendix B
Occupational Safety and Health Act of 1970

Public Law 91-596
91st Congress, S. 2193
December 29, 1970

An Act

84 STAT. 1590

To assure safe and healthful working conditions for working men and women; by authorizing enforcement of the standards developed under the Act; by assisting and encouraging the States in their efforts to assure safe and healthful working conditions; by providing for research, information, education, and training in the field of occupational safety and health; and for other purposes.

Be it enacted by the Senate and House of Representatives of the United States of America in Congress assembled, That this Act may be cited as the "Occupational Safety and Health Act of 1970".

Occupational Safety and Health Act of 1970.

CONGRESSIONAL FINDINGS AND PURPOSE

SEC. (2) The Congress finds that personal injuries and illnesses arising out of work situations impose a substantial burden upon, and are a hindrance to, interstate commerce in terms of lost production, wage loss, medical expenses, and disability compensation payments.

(b) The Congress declares it to be its purpose and policy, through the exercise of its powers to regulate commerce among the several States and with foreign nations and to provide for the general welfare, to assure so far as possible every working man and woman in the Nation safe and healthful working conditions and to preserve our human resources—

(1) by encouraging employers and employees in their efforts to reduce the number of occupational safety and health hazards at their places of employment, and to stimulate employers and employees to institute new and to perfect existing programs for providing safe and healthful working conditions;

(2) by providing that employers and employees have separate but dependent responsibilities and rights with respect to achieving safe and healthful working conditions;

(3) by authorizing the Secretary of Labor to set mandatory occupational safety and health standards applicable to businesses affecting interstate commerce, and by creating an Occupational Safety and Health Review Commission for carrying out adjudicatory functions under the Act;

(4) by building upon advances already made through employer and employee initiative for providing safe and healthful working conditions;

(5) by providing for research in the field of occupational safety and health, including the psychological factors involved, and by developing innovative methods, techniques, and approaches for dealing with occupational safety and health problems;

(6) by exploring ways to discover latent diseases, establishing causal connections between diseases and work in environmental conditions, and conducting other research relating to health problems, in recognition of the fact that occupational health standards present problems often different from those involved in occupational safety;

(7) by providing medical criteria which will assure insofar as practicable that no employee will suffer diminished health, functional capacity, or life expectancy as a result of his work experience;

(8) by providing for training programs to increase the number and competence of personnel engaged in the field of occupational safety and health;

Pub. Law 91-596 - 2 - December 29, 1970
84 STAT. 1591

(9) by providing for the development and promulgation of occupational safety and health standards;

(10) by providing an effective enforcement program which shall include a prohibition against giving advance notice of any inspection and sanctions for any individual violating this prohibition;

(11) by encouraging the States to assume the fullest responsibility for the administration and enforcement of their occupational safety and health laws by providing grants to the States to assist in identifying their needs and responsibilities in the area of occupational safety and health, to develop plans in accordance with the provisions of this Act, to improve the administration and enforcement of State occupational safety and health laws, and to conduct experimental and demonstration projects in connection therewith;

(12) by providing for appropriate reporting procedures with respect to occupational safety and health which procedures will help achieve the objectives of this Act and accurately describe the nature of the occupational safety and health problem;

(13) by encouraging joint labor-management efforts to reduce injuries and disease arising out of employment.

DEFINITIONS

SEC. 3. For the purposes of this Act—

(1) The term "Secretary" mean the Secretary of Labor.

(2) The term "Commission" means the Occupational Safety and Health Review Commission established under this Act.

(3) The term "commerce" means trade, traffic, commerce, transportation, or communication among the several States, or between a State and any place outside thereof, or within the District of Columbia, or a possession of the United States (other than the Trust Territory of the Pacific Islands), or between points in the same State but through a point outside thereof.

(4) The term "person" means one or more individuals, partnerships, associations, corporations, business trusts, legal representatives, or any organized group of persons.

(5) The term "employer" means a person engaged in a business affecting commerce who has employees, but does not include the United States or any State or political subdivision of a State.

(6) The term "employee" means an employee of an employer who is employed in a business of his employer which affects commerce.

(7) The term "State" includes a State of the United States, the District of Columbia, Puerto Rico, the Virgin Islands, American Samoa, Guam, and the Trust Territory of the Pacific Islands.

(8) The term "occupational safety and health standard" means a standard which requires conditions, or the adoption or use of one or more practices, means, methods, operations, or processes, reasonably necessary or appropriate to provide safe or healthful employment and places of employment.

(9) The term "national consensus standard" means any occupational safety and health standard or modification thereof which (1), has been adopted and promulgated by a nationally recognized standards-producing organization under procedures whereby it can be determined by the Secretary that persons interested

and affected by the scope or provisions of the standard have reached substantial agreement on its adoption, (2) was formulated in a manner which afforded an opportunity for diverse views to be considered and (3) has been designated as such a standard by the Secretary, after consultation with other appropriate Federal agencies.

(10) The term "established Federal standard" means any operative occupational safety and health standard established by any agency of the United States and presently in effect, or contained in any Act of Congress in force on the date of enactment of this Act.

(11) The term "Committee" means the National Advisory Committee on Occupational Safety and Health established under this Act.

(12) The term "Director" means the Director of the National Institute for Occupational Safety and Health.

(13) The term "Institute" means the National Institute for Occupational Safety and Health established under this Act.

(14) The term "Workmen's Compensation Commission" means the National Commission on State Workmen's Compensation Laws established under this Act.

APPLICABILITY OF THIS ACT

SEC. 4. (a) This Act shall apply with respect to employment performed in a workplace in a State, the District of Columbia, the Commonwealth of Puerto Rico, the Virgin Islands, American Samoa, Guam, the Trust Territory of the Pacific Islands, Wake Island, Outer Continental Shelf lands defined in the Outer Continental Shelf Lands Act, Johnston Island, and the Canal Zone. The Secretary of the Interior shall, by regulation, provide for judicial enforcement of this Act by the courts established for areas in which there are no United States district courts having jurisdiction.

67 Stat. 462. 43 USC 1331 note.

(b)(1) Nothing in this Act shall apply to working conditions of employees with respect to which other Federal agencies, and State agencies acting under section 274 of the Atomic Energy Act of 1954, as amended (42 U.S.C. 2021), exercise statutory authority to prescribe or enforce standards or regulations affecting occupational safety or health.

73 Stat. 688.

(2) The safety and health standards promulgated under the Act of June 30, 1936, commonly known as the Walsh-Healey Act (41 U.S.C. 35 et seq.), the Service Contract Act of 1965 (41 U.S.C. 351 et seq.), Public Law 91-54, Act of August 9, 1969 (40 U.S.C. 333), Public Law 85-742, Act of August 23, 1958 (33 U.S.C. 941), and the National Foundation on Arts and Humanities Act (20 U.S.C. 951 et seq.) are superseded on the effective date of corresponding standards, promulgated under this Act, which are determined by the Secretary to be more effective. Standards issued under the laws listed in this paragraph and in effect on or after the effective date of this Act shall be deemed to be occupational safety and health standards issued under this Act, as well as under such other Acts.

49 Stat. 2036. 79 Stat. 1034. 83 Stat. 96. 72 Stat. 835. 79 Stat. 845; Ante, p. 443.

(3) The Secretary shall, within three years after the effective date of this Act, report to the Congress his recommendations for legislation to avoid unnecessary duplication and to achieve coordination between this Act and other Federal laws.

Report to Congress.

(4) Nothing in this Act shall be construed to supersede or in any manner affect any workmen's compensation law or to enlarge or diminish or affect in any other manner the common law or statutory rights, duties, or liabilities of employers and employees under any law with respect to injuries, diseases, or death of employees arising out of, or in the course of, employment.

DUTIES

SEC. 5. (a) Each employer—

(1) shall furnish to each of his employees employment and a place of employment which are free from recognized hazards that are causing or are likely to cause death or serious physical harm to his employees;

(2) shall comply with occupational safety and health standards promulgated under this Act.

(b) Each employee shall comply with occupational safety and health standards and all rules, regulations, and orders issued pursuant to this Act which are applicable to his own actions and conduct.

OCCUPATIONAL SAFETY AND HEALTH STANDARDS

80 Stat. 381; 81 Stat. 195. 5 USC 500.

SEC. 6. (a) Without regard to chapter 5 of title 5, United States Code, or to the other subsections of this section, the Secretary shall, as soon as practicable during the period beginning with the effective date of this Act and ending two years after such date, by rule promulgate as an occupational safety or health standard any national consensus standard, and any established Federal standard, unless he determines that the promulgation of such a standard would not result in improved safety or health for specifically designated employees. In the event of conflict among any such standards, the Secretary shall promulgate the standard which assures the greatest protection of the safety or health of the affected employees.

(b) The Secretary may by rule promulgate, modify, or revoke any occupational safety or health standard in the following manner:

Advisory committee, recommendations.

(1) Whenever the Secretary, upon the basis of information submitted to him in writing by an interested person, a representative of any organization of employers or employees, a nationally recognized standards-producing organization, the Secretary of Health, Education, and Welfare, the National Institute for Occupational Safety and Health, or a State or political subdivision, or on the basis of information developed by the Secretary or otherwise available to him, determines that a rule should be promulgated in order to serve the objectives of this Act, the Secretary may request the recommendations of an advisory committee appointed under section 7 of this Act. The Secretary shall provide such an advisory committee with any proposals of his own or of the Secretary of Health, Education, and Welfare, together with all pertinent factual information developed by the Secretary or the Secretary of Health, Education, and Welfare, or otherwise available, including the results of research, demonstrations, and experiments. An advisory committee shall submit to the Secretary its recommendations regarding the rule to be promulgated within ninety days from the date of its appointment or within such longer or shorter period as may be prescribed by the Secretary, but in no event for a period which is longer than two hundred and seventy days.

(2) The Secretary shall publish a proposed rule promulgating, modifying, or revoking an occupational safety or health standard in the Federal Register and shall afford interested persons a period of thirty days after publication to submit written data or comments. Where an advisory committee is appointed and the Secretary determines that a rule should be issued, he shall publish the proposed rule within sixty days after the submission of the advisory committee's recommendations or the expiration of the period prescribed by the Secretary for such submission.

Publication in Federal Register.

(3) On or before the last day of the period provided for the submission of written data or comments under paragraph (2), any interested person may file with the Secretary written objections to the proposed rule, stating the grounds therefor and requesting a public hearing on such objections. Within thirty days after the last day for filing such objections, the Secretary shall publish in the Federal Register a notice specifying the occupational safety or health standard to which objections have been filed and a hearing requested, and specifying a time and place for such hearing.

Hearing, notice.

Publication in Federal Register.

(4) Within sixty days after the expiration of the period provided for the submission of written data or comments under paragraph (2), or within sixty days after the completion of any hearing held under paragraph (3), the Secretary shall issue a rule promulgating, modifying, or revoking an occupational safety or health standard or make a determination that a rule should not be issued. Such a rule may contain a provision delaying its effective date for such period (not in excess of ninety days) as the Secretary determines may be necessary to insure that affected employers and employees will be informed of the existence of the standard and of its terms and that employers affected are given an opportunity to familiarize themselves and their employees with the existence of the requirements of the standard.

(5) The Secretary, in promulgating standards dealing with toxic materials or harmful physical agents under this subsection, shall set the standard which most adequately assures, to the extent feasible, on the basis of the best available evidence, that no employee will suffer material impairment of health or functional capacity even if such employee has regular exposure to the hazard dealt with by such standard for the period of his working life. Development of standards under this subsection shall be based upon research, demonstrations, experiments, and such other information as may be appropriate. In addition to the attainment of the highest degree of health and safety protection for the employee, other considerations shall be the latest available scientific data in the field, the feasibility of the standards, and experience gained under this and other health and safety laws. Whenever practicable, the standard promulgated shall be expressed in terms of objective criteria and of the performance desired.

Toxic materials.

(6)(A) Any employer may apply to the Secretary for a temporary order granting a variance from a standard or any provision thereof promulgated under this section. Such temporary order shall be granted only if the employer files an application which meets the requirements of clause (B) and establishes that (i) he is unable to comply with a standard by its effective date because of unavailability of professional or technical personnel or of materials and equipment needed to come into compliance with the standard or because necessary construction or alteration of facilities cannot be completed by the effective date, (ii) he is taking all available steps to safeguard his employees against the hazards covered by the standard, and (iii) he has an effective program for coming into compliance with the standard as quickly as

Temporary variance order.

practicable. Any temporary order issued under this paragraph shall prescribe the practices, means, methods, operations, and processes which the employer must adopt and use while the order is in effect and state in detail his program for coming into compliance with the standard. Such a temporary order may be granted only after notice to employees and an opportunity for a hearing: *Provided*, That the Secretary may issue one interim order to be effective until a decision is made on the basis of the hearing. No temporary order may be in effect for longer than the period needed by the employer to achieve compliance with the standard or one year, whichever is shorter, except that such an order may be renewed not more than twice (I) so long as the requirements of this paragraph are met and (II) if an application for renewal is filed at least 90 days prior to the expiration date of the order. No interim renewal of an order may remain in effect for longer than 180 days.

Notice, hearing.

Renewal.

Time limitation.

(B) An application for a temporary order under this paragraph (6) shall contain:

(i) a specification of the standard or portion thereof from which the employer seeks a variance,

(ii) a representation by the employer, supported by representations from qualified persons having firsthand knowledge of the facts represented, that he is unable to comply with the standard or portion thereof and a detailed statement of the reasons therefor,

(iii) a statement of the steps he has taken and will take (with specific dates) to protect employees against the hazard covered by the standard,

(iv) a statement of when he expects to be able to comply with the standard and what steps he has taken and what steps he will take (with dates specified) to come into compliance with the standard, and

(v) a certification that he has informed his employees of the application by giving a copy thereof to their authorized representative, posting a statement giving a summary of the application and specifying where a copy may be examined at the place or places where notices to employees are normally posted, and by other appropriate means.

A description of how employees have been informed shall be contained in the certification. The information to employees shall also inform them of their right to petition the Secretary for a hearing.

(C) The Secretary is authorized to grant a variance from any standard or portion thereof whenever he determines, or the Secretary of Health, Education, and Welfare certifies, that such variance is necessary to permit an employer to participate in an experiment approved by him or the Secretary of Health, Education, and Welfare designed to demonstrate or validate new and improved techniques to safeguard the health or safety of workers.

Labels, etc.

(7) Any standard promulgated under this subsection shall prescribe the use of labels or other appropriate forms of warning as are necessary to insure that employees are apprised of all hazards to which they are exposed, relevant symptoms and appropriate emergency treatment, and proper conditions and precautions of safe use or exposure. Where appropriate, such standard shall also prescribe suitable protective equipment and control or technological procedures to be used in connection with such hazards and shall provide for monitoring or measuring employee exposure at such locations and intervals, and in such manner as may be necessary for the protection of employees. In

Protective equipment, etc.

addition, where appropriate, any such standard shall prescribe the type and frequency of medical examinations or other tests which shall be made available, by the employer or at his cost, to employees exposed to such hazards in order to most effectively determine whether the health of such employees is adversely affected by such exposure. In the event such medical examinations are in the nature of research, as determined by the Secretary of Health, Education, and Welfare, such examinations may be furnished at the expense of the Secretary of Health, Education, and Welfare. The results of such examinations or tests shall be furnished only to the Secretary or the Secretary of Health, Education, and Welfare, and, at the request of the employee, to his physician. The Secretary, in consultation with the Secretary of Health, Education, and Welfare, may by rule promulgated pursuant to section 553 of title 5, United States Code, make appropriate modifications in the foregoing requirements relating to the use of labels or other forms of warning, monitoring or measuring, and medical examinations, as may be warranted by experience, information, or medical or technological developments acquired subsequent to the promulgation of the relevant standard.

Medical examinations.

80 Stat. 383.

(8) Whenever a rule promulgated by the Secretary differs substantially from an existing national consensus standard, the Secretary shall, at the same time, publish in the Federal Register a statement of the reasons why the rule as adopted will better effectuate the purposes of this Act than the national consensus standard.

Publication in Federal Register.

(c)(1) The Secretary shall provide, without regard to the requirements of chapter 5, title 5, United States Code, for an emergency temporary standard to take immediate effect upon publication in the Federal Register if he determines (A) that employees are exposed to grave danger from exposure to substances or agents determined to be toxic or physically harmful or from new hazards, and (B) that such emergency standard is necessary to protect employees from such danger.

Temporary standard. Publication in Federal Register. 80 Stat. 381; 81 Stat. 195. 5 USC 500.

(2) Such standard shall be effective until superseded by a standard promulgated in accordance with the procedures prescribed in paragraph (3) of this subsection.

Time limitation.

(3) Upon publication of such standard in the Federal Register the Secretary shall commence a proceeding in accordance with section 6(b) of this Act, and the standard as published shall also serve as a proposed rule for the proceeding. The Secretary shall promulgate a standard under this paragraph no later than six months after publication of the emergency standard as provided in paragraph (2) of this subsection.

(d) Any affected employer may apply to the Secretary for a rule or order for a variance from a standard promulgated under this section. Affected employees shall be given notice of each such application and an opportunity to participate in a hearing. The Secretary shall issue such rule or order if he determines on the record, after opportunity for an inspection where appropriate and a hearing, that the proponent of the variance has demonstrated by a preponderance of the evidence that the conditions, practices, means, methods, operations, or processes used or proposed to be used by an employer will provide employment and places of employment to his employees which are as safe and healthful as those which would prevail if he complied with the standard. The rule or order so issued shall prescribe the conditions the employer must maintain, and the practices, means, methods, operations, and processes which he must adopt and utilize to the extent they

Variance rule.

differ from the standard in question. Such a rule or order may be modified or revoked upon application by an employer, employees, or by the Secretary on his own motion, in the manner prescribed for its issuance under this subsection at any time after six months from its issuance.

Publication in Federal Register.

(e) Whenever the Secretary promulgates any standard, makes any rule, order, or decision, grants any exemption or extension of time, or compromises, mitigates, or settles any penalty assessed under this Act, he shall include a statement of the reasons for such action, which shall be published in the Federal Register.

Petition for judicial review.

(f) Any person who may be adversely affected by a standard issued under this section may at any time prior to the sixtieth day after such standard is promulgated file a petition challenging the validity of such standard with the United States court of appeals for the circuit wherein such person resides or has his principal place of business, for a judicial review of such standard. A copy of the petition shall be forthwith transmitted by the clerk of the court to the Secretary. The filing of such petition shall not, unless otherwise ordered by the court, operate as a stay of the standard. The determinations of the Secretary shall be conclusive if supported by substantial evidence in the record considered as a whole.

(g) In determining the priority for establishing standards under this section, the Secretary shall give due regard to the urgency of the need for mandatory safety and health standards for particular industries, trades, crafts, occupations, businesses, workplaces or work environments. The Secretary shall also give due regard to the recommendations of the Secretary of Health, Education, and Welfare regarding the need for mandatory standards in determining the priority for establishing such standards.

ADVISORY COMMITTEES; ADMINISTRATION

Establishment; membership.

SEC. 7. (a)(1) There is hereby established a National Advisory Committee on Occupational Safety and Health consisting of twelve members appointed by the Secretary, four of whom are to be designated by the Secretary of Health, Education, and Welfare, without regard to the provisions of title 5, United States Code, governing appointments in the competitive service, and composed of representatives of manangement, labor, occupational safety and occupational health professions, and of the public. The Secretary shall designate one of the public members as Chairman. The members shall be selected upon the basis of their experience and competence in the field of occupational safety and health.

80 Stat. 378. 5 USC 101.

(2) The Committee shall advise, consult with, and make recommendations to the Secretary and the Secretary of Health, Education, and Welfare on matters relating to the administration of the Act. The Committee shall hold no fewer than two meetings during each calendar year. All meetings of the Committee shall be open to the public and a transcript shall be kept and made available for public inspection.

Public transcript.

(3) The members of the Committee shall be compensated in accordance with the provisions of section 3109 of title 5, United States Code.

80 Stat. 416.

(4) The Secretary shall furnish to the Committee an executive secretary and such secretarial, clerical, and other services as are deemed necessary to the conduct of its business.

(b) An advisory committee may be appointed by the Secretary to assist him in his standard-setting functions under section 6 of this Act. Each such committee shall consist of not more than fifteen members and shall include as a member one or more designees of the Secretary of Health, Education, and Welfare, and shall include among its members an equal number of persons qualified by experience and affiliation to present the viewpoint of the employers involved, and of persons similarly qualified to present the viewpoint of the workers involved, as well as one or more represent..tives of health and safety agencies of the States. An advisory committee may also include such other persons as the Secretary may appoint who are qualified by knowledge and experience to make a useful contribution to the work of such committee, including one or more representatives of professional organizations of technicians or professionals specializing in occupational safety or health, and one or more representatives of nationally recognized standards-producing organizations, but the number of persons so appointed to any such advisory committee shall not exceed the number appointed to such committee as representatives of Federal and State agencies. Persons appointed to advisory committees from private life shall be compensated in the same manner as consultants or experts under section 3109 of title 5, United States Code. The Secretary shall pay to any State which is the employer of a member of such a committee who is a representative of the health or safety agency of that State, reimbursement sufficient to cover the actual cost to the State resulting from such representative's membership on such committee. Any meeting of such committee shall be open to the public and an accurate record shall be kept and made available to the public. No member of such committee (other than representatives of employers and employees) shall have an economic interest in any proposed rule.

80 Stat. 416.

Recordkeeping.

(c) In carrying out his responsibilities under this Act, the Secretary is authorized to—

(1) use, with the consent of any Federal agency, the services, facilities, and personnel of such agency, with or without reimbursement, and with the consent of any State or political subdivision thereof, accept and use the services, facilities, and personnel of any agency of such State or subdivision with reimbursement; and

(2) employ experts and consultants or organizations thereof as authorized by section 3109 of title 5, United States Code, except that contracts for such employment may be renewed annually; compensate individuals so employed at rates not in excess of the rate specified at the time of service for grade GS-18 under section 5332 of title 5, United States Code, including traveltime, and allow them while away from their homes or regular places of business, travel expenses (including per diem in lieu of subsistence) as authorized by section 5703 of title 5, United States Code, for persons in the Government service employed intermittently, while so employed.

Ante, p. 198-1.

80 Stat. 499; 83 Stat. 190.

INSPECTIONS, INVESTIGATIONS, AND RECORDKEEPING

Sec. 8. (a) In order to carry out the purposes of this Act, the Secretary, upon presenting appropriate credentials to the owner, operator, or agent in charge, is authorized—

(1) to enter without delay and at reasonable times any factory, plant, establishment, construction site, or other area, workplace or environment where work is performed by an employee of an employer; and

(2) to inspect and investigate during regular working hours and at other reasonable times, and within reasonable limits and in a reasonable manner, any such place of employment and all pertinent conditions, structures, machines, apparatus, devices, equipment, and materials therein, and to question privately any such employer, owner, operator, agent or employee.

Subpoena power.

(b) In making his inspections and investigations under this Act the Secretary may require the attendance and testimony of witnesses and the production of evidence under oath. Witnesses shall be paid the same fees and mileage that are paid witnesses in the courts of the United States. In case of a contumacy, failure, or refusal of any person to obey such an order, any district court of the United States or the United States courts of any territory or possession, within the jurisdiction of which such person is found, or resides or transacts business, upon the application by the Secretary, shall have jurisdiction to issue to such person an order requiring such person to appear to produce evidence if, as, and when so ordered, and to give testimony relating to the matter under investigation or in question, and any failure to obey such order of the court may be punished by said court as a contempt thereof.

Recordkeeping.

(c)(1) Each employer shall make, keep and preserve, and make available to the Secretary or the Secretary of Health, Education, and Welfare, such records regarding his activities relating to this Act as the Secretary, in cooperation with the Secretary of Health, Education, and Welfare, may prescribe by regulation as necessary or appropriate for the enforcement of this Act or for developing information regarding the causes and prevention of occupational accidents and illnesses. In order to carry out the provisions of this paragraph such regulations may include provisions requiring employers to conduct periodic inspections. The Secretary shall also issue regulations requiring that employers, through posting of notices or other appropriate means, keep their employees informed of their protections and obligations under this Act, including the provisions of applicable standards.

Work-related deaths, etc.; reports.

(2) The Secretary, in cooperation with the Secretary of Health, Education, and Welfare, shall prescribe regulations requiring employers to maintain accurate records of, and to make periodic reports on, work-related deaths, injuries and illnesses other than minor injuries requiring only first aid treatment and which do not involve medical treatment, loss of consciousness, restriction of work or motion, or transfer to another job.

(3) The Secretary, in cooperation with the Secretary of Health, Education, and Welfare, shall issue regulations requiring employers to maintain accurate records of employee exposures to potentially toxic materials or harmful physical agents which are required to be monitored or measured under section 6. Such regulations shall provide employees or their representatives with an opportunity to observe such monitoring or measuring, and to have access to the records thereof. Such regulations shall also make appropriate provision for each employee or former employee to have access to such records as will indicate his own exposure to toxic materials or harmful physical agents. Each employer shall promptly notify any employee who has been or is being exposed to toxic materials or harmful physical agents in concentrations or at levels which exceed those prescribed by an applicable occupational safety and health standard promulgated under section 6, and shall inform any employee who is being thus exposed of the corrective action being taken.

(d) Any information obtained by the Secretary, the Secretary of Health, Education, and Welfare, or a State agency under this Act shall be obtained with a minimum burden upon employers, especially those operating small businesses. Unnecessary duplication of efforts in obtaining information shall be reduced to the maximum extent feasible.

(e) Subject to regulations issued by the Secretary, a representative of the employer and a representative authorized by his employees shall be given an opportunity to accompany the Secretary or his authorized representative during ~~the physic~~al inspection of any workplace under subsection (a) for the purpose of aiding such inspection. Where there is no authorized employee representative, the Secretary or his authorized representative shall consult with a reasonable number of employees concerning matters of health and safety in the workplace.

(f)(1) Any employees or representative of employees who believe that a violation of a safety or health standard exists that threatens physical harm, or that an imminent danger exists, may request an inspection by giving notice to the Secretary or his authorized representative of such violation or danger. Any such notice shall be reduced to writing, shall set forth with reasonable particularity the grounds for the notice, and shall be signed by the employees or representative of employees, and a copy shall be provided the employer or his agent no later than at the time of inspection, except that, upon the request of the person giving such notice, his name and the names of individual employees referred to therein shall not appear in such copy or on any record published, released, or made available pursuant to subsection (g) of this section. If upon receipt of such notification the Secretary determines there are reasonable grounds to believe that such violation or danger exists, he shall make a special inspection in accordance with the provisions of this section as soon as practicable, to determine if such violation or danger exists. If the Secretary determines there are no reasonable grounds to believe that a violation or danger exists he shall notify the employees or representative of the employees in writing of such determination.

(2) Prior to or during any inspection of a workplace, any employees or representative of employees employed in such workplace may notify the Secretary or any representative of the Secretary responsible for conducting the inspection, in writing, of any violation of this Act which they have reason to believe exists in such workplace. The Secretary shall, by regulation, establish procedures for informal review of any refusal by a representative of the Secretary to issue a citation with respect to any such alleged violation and shall furnish the employees or representative of employees requesting such review a written statement of the reasons for the Secretary's final disposition of the case.

Reports, publication.

(g)(1) The Secretary and Secretary of Health, Education, and Welfare are authorized to compile, analyze, and publish, either in summary or detailed form, all reports or information obtained under this section.

Rules and regulations.

(2) The Secretary and the Secretary of Health, Education, and Welfare shall each prescribe such rules and regulations as he may deem necessary to carry out their responsibilities under this Act, including rules and regulations dealing with the inspection of an employer's establishment.

CITATIONS

SEC. 9. (a) If, upon inspection or investigation, the Secretary or his authorized representative believes that an employer has violated a requirement of section 5 of this Act, of any standard, rule or order promulgated pursuant to section 6 of this Act, or of any regulations prescribed pursuant to this Act, he shall with reasonable promptness issue a citation to the employer. Each citation shall be in writing and shall describe with particularity the nature of the violation, including a reference to the provision of the Act, standard, rule, regulation, or order alleged to have been violated. In addition, the citation shall fix a reasonable time for the abatement of the violation. The Secretary may prescribe procedures for the issuance of a notice in lieu of a citation with respect to de minimis violations which have no direct or immediate relationship to safety or health.

(b) Each citation issued under this section, or a copy or copies thereof, shall be prominently posted, as prescribed in regulations issued by the Secretary, at or near each place a violation referred to in the citation occurred.

Limitation.

(c) No citation may be issued under this section after the expiration of six months following the occurrence of any violation.

PROCEDURE FOR ENFORCEMENT

SEC. 10. (a) If, after an inspection or investigation, the Secretary issues a citation under section 9(a), he shall, within a reasonable time after the termination of such inspection or investigation, notify the employer by certified mail of the penalty, if any, proposed to be assessed under section 17 and that the employer has fifteen working days within which to notify the Secretary that he wishes to contest the citation or proposed assessment of penalty. If, within fifteen working days from the receipt of the notice issued by the Secretary the employer fails to notify the Secretary that he intends to contest the citation or proposed assessment of penalty, and no notice is filed by any employee or representative of employees under subsection (c) within such time, the citation and the assessment, as proposed, shall be deemed a final order of the Commission and not subject to review by any court or agency.

(b) If the Secretary has reason to believe that an employer has failed to correct a violation for which a citation has been issued within the period permitted for its correction (which period shall not begin to run until the entry of a final order by the Commission in the case of any review proceedings under this section initiated by the employer in good faith and not solely for delay or avoidance of penalties), the Secretary shall notify the employer by certified mail of such failure and of the penalty proposed to be assessed under section 17 by reason of such failure, and that the employer has fifteen working days within which to notify the Secretary that he wishes to contest the Secretary's notification or the proposed assessment of penalty. If, within fifteen working days from the receipt of notification issued by the Secretary, the employer fails to notify the Secretary that he intends to contest the notification or proposed assessment of penalty, the notification and assessment, as proposed, shall be deemed a final order of the Commission and not subject to review by any court or agency.

(c) If an employer notifies the Secretary that he intends to contest a citation issued under section 9(a) or notification issued under subsection (a) or (b) of this section, or if, within fifteen working days of the issuance of a citation under section 9(a), any employee or representative of employees files a notice with the Secretary alleging that the period of time fixed in the citation for the abatement of the violation is unreasonable, the Secretary shall immediately advise the Commission of such notification, and the Commission shall afford an opportunity for a hearing (in accordance with section 554 of title 5, United States Code, but without regard to subsection (a)(3) of such section). The Commission shall thereafter issue an order, based on findings of fact, affirming, modifying, or vacating the Secretary's citation or proposed penalty, or directing other appropriate relief, and such order shall become final thirty days after its issuance. Upon a showing by an employer of a good faith effort to comply with the abatement requirements of a citation, and that abatement has not been completed because of factors beyond his reasonable control, the Secretary, after an opportunity for a hearing as provided in this subsection, shall issue an order affirming or modifying the abatement requirements in such citation. The rules of procedure prescribed by the Commission shall provide affected employees or representatives of affected employees an opportunity to participate as parties to hearings under this subsection.

80 Stat. 384.

JUDICIAL REVIEW

SEC. 11. (a) Any person adversely affected or aggrieved by an order of the Commission issued under subsection (c) of section 10 may obtain a review of such order in any United States court of appeals for the circuit in which the violation is alleged to have occurred or where the employer has its principal office, or in the Court of Appeals for the District of Columbia Circuit, by filing in such court within sixty days following the issuance of such order a written petition praying that the order be modified or set aside. A copy of such petition shall be forthwith transmitted by the clerk of the court to the Commission and to the other parties, and thereupon the Commission shall file in the court the record in the proceeding as provided in section 2112 of title 28, United States Code. Upon such filing, the court shall have jurisdiction of the proceeding and of the question determined therein, and shall have power to grant such temporary relief or restraining order as it deems just and proper, and to make and enter upon the pleadings, testimony, and proceedings set forth in such record a decree affirming, modifying, or setting aside in whole or in part, the order of the Commission and enforcing the same to the extent that such order is affirmed or modified. The commencement of proceedings under this subsection shall not, unless ordered by the court, operate as a stay of the order of the Commission. No objection that has not been urged before the Commission shall be considered by the court, unless the failure or neglect to urge such objection shall be excused because of extraordinary circumstances. The findings of the Commission with respect to questions of fact, if supported by substantial evidence on the record considered as a whole, shall be conclusive. If any party shall apply to the court for leave to adduce additional evidence and shall show to the satisfaction of the court that such additional evidence is material and that there were reasonable grounds for the failure to adduce such evidence in the hearing before the Commission, the court may order such additional evidence to be taken before the Commission and to be made a part of the record. The Commission may modify its findings as to the facts, or make new findings, by reason of additional evidence so taken and filed, and it shall file such modified or new findings, which findings with respect to questions of fact, if supported by substantial evi-

72 Stat. 941;
80 Stat. 1323.

dence on the record considered as a whole, shall be conclusive, and its recommendations, if any, for the modification or setting aside of its original order. Upon the filing of the record with it, the jurisdiction of the court shall be exclusive and its judgment and decree shall be final, except that the same shall be subject to review by the Supreme Court of the United States, as provided in section 1254 of title 28, United States Code. Petitions filed under this subsection shall be heard expeditiously.

62 Stat. 928.

(b) The Secretary may also obtain review or enforcement of any final order of the Commission by filing a petition for such relief in the United States court of appeals for the circuit in which the alleged violation occurred or in which the employer has its principal office, and the provisions of subsection (a) shall govern such proceedings to the extent applicable. If no petition for review, as provided in subsection (a), is filed within sixty days after service of the Commission's order, the Commission's findings of fact and order shall be conclusive in connection with any petition for enforcement which is filed by the Secretary after the expiration of such sixty-day period. In any such case, as well as in the case of a noncontested citation or notification by the Secretary which has become a final order of the Commission under subsection (a) or (b) of section 10, the clerk of the court, unless otherwise ordered by the court, shall forthwith enter a decree enforcing the order and shall transmit a copy of such decree to the Secretary and the employer named in the petition. In any contempt proceeding brought to enforce a decree of a court of appeals entered pursuant to this subsection or subsection (a), the court of appeals may assess the penalties provided in section 17, in addition to invoking any other available remedies.

(c) (1) No person shall discharge or in any manner discriminate against any employee because such employee has filed any complaint or instituted or caused to be instituted any proceeding under or related to this Act or has testified or is about to testify in any such proceeding or because of the exercise by such employee on behalf of himself or others of any right afforded by this Act.

(2) Any employee who believes that he has been discharged or otherwise discriminated against by any person in violation of this subsection may, within thirty days after such violation occurs, file a complaint with the Secretary alleging such discrimination. Upon receipt of such complaint, the Secretary shall cause such investigation to be made as he deems appropriate. If upon such investigation, the Secretary determines that the provisions of this subsection have been violated, he shall bring an action in any appropriate United States district court against such person. In any such action the United States district courts shall have jurisdiction, for cause shown to restrain violations of paragraph (1) of this subsection and order all appropriate relief including rehiring or reinstatement of the employee to his former position with back pay.

(3) Within 90 days of the receipt of a complaint filed under the subsection the Secretary shall notify the complainant of his determination under paragraph 2 of this subsection.

THE OCCUPATIONAL SAFETY AND HEALTH REVIEW COMMISSION

Establishment; membership.

SEC. 12. (a) The Occupational Safety and Health Review Commission is hereby established. The Commission shall be composed of three members who shall be appointed by the President, by and with the advice and consent of the Senate, from among persons who by reason of training, education, or experience are qualified to carry out the functions of the Commission under this Act. The President shall designate one of the members of the Commission to serve as Chairman.

(b) The terms of members of the Commission shall be six years except that (1) the members of the Commission first taking office shall serve, as designated by the President at the time of appointment, one for a term of two years, one for a term of four years, and one for a term of six years, and (2) a vacancy caused by the death, resignation, or removal of a member prior to the expiration of the term for which he was appointed shall be filled only for the remainder of such unexpired term. A member of the Commission may be removed by the President for inefficiency, neglect of duty, or malfeasance in office.

Terms.

(c)(1) Section 5314 of title 5, United States Code, is amended by adding at the end thereof the following new paragraph:

80 Stat. 460.

"(57) Chairman, Occupational Safety and Health Review Commission."

(2) Section 5315 of title 5, United States Code, is amended by adding at the end thereof the following new paragraph:

Ante, p. 776.

"(94) Members, Occupational Safety and Health Review Commission."

(d) The principal office of the Commission shall be in the District of Columbia. Whenever the Commission deems that the convenience of the public or of the parties may be promoted, or delay or expense may be minimized, it may hold hearings or conduct other proceedings at any other place.

Location.

(e) The Chairman shall be responsible on behalf of the Commission for the administrative operations of the Commission and shall appoint such hearing examiners and other employees as he deems necessary to assist in the performance of the Commission's functions and to fix their compensation in accordance with the provisions of chapter 51 and subchapter III of chapter 53 of title 5, United States Code, relating to classification and General Schedule pay rates: *Provided*, That assignment, removal and compensation of hearing examiners shall be in accordance with sections 3105, 3344, 5362, and 7521 of title 5, United States Code.

5 USC 5101, 5331.

Ante, p. 198-1.

(f) For the purpose of carrying out its functions under this Act, two members of the Commission shall constitute a quorum and official action can be taken only on the affirmative vote of at least two members.

Quorum.

(g) Every official act of the Commission shall be entered of record, and its hearings and records shall be open to the public. The Commission is authorized to make such rules as are necessary for the orderly transaction of its proceedings. Unless the Commission has adopted a different rule, its proceedings shall be in accordance with the Federal Rules of Civil Procedure.

Public records.

28 USC app.

(h) The Commission may order testimony to be taken by deposition in any proceedings pending before it at any state of such proceeding. Any person may be compelled to appear and depose, and to produce books, papers, or documents, in the same manner as witnesses may be compelled to appear and testify and produce like documentary evidence before the Commission. Witnesses whose depositions are taken under this subsection, and the persons taking such depositions, shall be entitled to the same fees as are paid for like services in the courts of the United States.

(i) For the purpose of any proceeding before the Commission, the provisions of section 11 of the National Labor Relations Act (29 U.S.C. 161) are hereby made applicable to the jurisdiction and powers of the Commission.

61 Stat. 150; Ante, p. 930.

Report.

(j) A hearing examiner appointed by the Commission shall hear, and make a determination upon, any proceeding instituted before the Commission and any motion in connection therewith, assigned to such hearing examiner by the Chairman of the Commission, and shall make a report of any such determination which constitutes his final disposition of the proceedings. The report of the hearing examiner shall become the final order of the Commission within thirty days after such report by the hearing examiner, unless within such period any Commission member has directed that such report shall be reviewed by the Commission.

(k) Except as otherwise provided in this Act, the hearing examiners shall be subject to the laws governing employees in the classified civil service, except that appointments shall be made without regard to section 5108 of title 5, United States Code. Each hearing examiner shall receive compensation at a rate not less than that prescribed for GS-16 under section 5332 of title 5, United States Code.

80 Stat. 453.

Ante, p. 198-1.

PROCEDURES TO COUNTERACT IMMINENT DANGERS

SEC. 13. (a) The United States district courts shall have jurisdiction, upon petition of the Secretary, to restrain any conditions or practices in any place of employment which are such that a danger exists which could reasonably be expected to cause death or serious physical harm immediately or before the imminence of such danger can be eliminated through the enforcement procedures otherwise provided by this Act. Any order issued under this section may require such steps to be taken as may be necessary to avoid, correct, or remove such imminent danger and prohibit the employment or presence of any individual in locations or under conditions where such imminent danger exists, except individuals whose presence is necessary to avoid, correct, or remove such imminent danger or to maintain the capacity of a continuous process operation to resume normal operations without a complete cessation of operations, or where a cessation of operations is necessary, to permit such to be accomplished in a safe and orderly manner.

(b) Upon the filing of any such petition the district court shall have jurisdiction to grant such injunctive relief or temporary restraining order pending the outcome of an enforcement proceeding pursuant to this Act. The proceeding shall be as provided by Rule 65 of the Federal Rules, Civil Procedure, except that no temporary restraining order issued without notice shall be effective for a period longer than five days.

28 USC app.

(c) Whenever and as soon as an inspector concludes that conditions or practices described in subsection (a) exist in any place of employment, he shall inform the affected employees and employers of the danger and that he is recommending to the Secretary that relief be sought.

(d) If the Secretary arbitrarily or capriciously fails to seek relief under this section, any employee who may be injured by reason of such failure, or the representative of such employees, might bring an action against the Secretary in the United States district court for the district in which the imminent danger is alleged to exist or the employer has its principal office, or for the District of Columbia, for a writ of mandamus to compel the Secretary to seek such an order and for such further relief as may be appropriate.

REPRESENTATION IN CIVIL LITIGATION

SEC. 14. Except as provided in section 518(a) of title 28, United States Code, relating to litigation before the Supreme Court, the Solicitor of Labor may appear for and represent the Secretary in any civil litigation brought under this Act but all such litigation shall be subject to the direction and control of the Attorney General.

80 Stat. 613.

CONFIDENTIALITY OF TRADE SECRETS

SEC. 15. All information reported to or otherwise obtained by the Secretary or his representative in connection with any inspection or proceeding under this Act which contains or which might reveal a trade secret referred to in section 1905 of title 18 of the United States Code shall be considered confidential for the purpose of that section, except that such information may be disclosed to other officers or employees concerned with carrying out this Act or when relevant in any proceeding under this Act. In any such proceeding the Secretary, the Commission, or the court shall issue such orders as may be appropriate to protect the confidentiality of trade secrets.

62 Stat. 791.

VARIATIONS, TOLERANCES, AND EXEMPTIONS

SEC. 16. The Secretary, on the record, after notice and opportunity for a hearing may provide such reasonable limitations and may make such rules and regulations allowing reasonable variations, tolerances, and exemptions to and from any or all provisions of this Act as he may find necessary and proper to avoid serious impairment of the national defense. Such action shall not be in effect for more than six months without notification to affected employees and an opportunity being afforded for a hearing.

PENALTIES

SEC. 17. (a) Any employer who willfully or repeatedly violates the requirements of section 5 of this Act, any standard, rule, or order promulgated pursuant to section 6 of this Act, or regulations prescribed pursuant to this Act, may be assessed a civil penalty of not more than $10,000 for each violation.

(b) Any employer who has received a citation for a serious violation of the requirements of section 5 of this Act, of any standard, rule, or order promulgated pursuant to section 6 of this Act, or of any regulations prescribed pursuant to this Act, shall be assessed a civil penalty of up to $1,000 for each such violation.

(c) Any employer who has received a citation for a violation of the requirements of section 5 of this Act, of any standard, rule, or order promulgated pursuant to section 6 of this Act, or of regulations prescribed pursuant to this Act, and such violation is specifically determined not to be of a serious nature, may be assessed a civil penalty of up to $1,000 for each such violation.

(d) Any employer who fails to correct a violation for which a citation has been issued under section 9(a) within the period permitted for its correction (which period shall not begin to run until the date of the final order of the Commission in the case of any review proceeding under section 10 initiated by the employer in good faith and not solely for delay or avoidance of penalties), may be assessed a civil penalty of not more than $1,000 for each day during which such failure or violation continues.

(e) Any employer who willfully violates any standard, rule, or order promulgated pursuant to section 6 of this Act, or of any regulations prescribed pursuant to this Act, and that violation caused death to any employee, shall, upon conviction, be punished by a fine of not more than $10,000 or by imprisonment for not more than six months, or by both; except that if the conviction is for a violation committed after a first conviction of such person, punishment shall be by a fine of not more than $20,000 or by imprisonment for not more than one year, or by both.

(f) Any person who gives advance notice of any inspection to be conducted under this Act, without authority from the Secretary or his designees, shall, upon conviction, be punished by a fine of not more than $1,000 or by imprisonment for not more than six months, or by both.

(g) Whoever knowingly makes any false statement, representation, or certification in any application, record, report, plan, or other document filed or required to be maintained pursuant to this Act shall, upon conviction, be punished by a fine of not more than $10,000, or by imprisonment for not more than six months, or by both.

65 Stat. 721; 79 Stat. 234.

(h)(1) Section 1114 of title 18, United States Code, is hereby amended by striking out "designated by the Secretary of Health, Education, and Welfare to conduct investigations, or inspections under the Federal Food, Drug, and Cosmetic Act" and inserting in lieu thereof "or of the Department of Labor assigned to perform investigative, inspection, or law enforcement functions".

62 Stat. 756.

(2) Notwithstanding the provisions of sections 1111 and 1114 of title 18, United States Code, whoever, in violation of the provisions of section 1114 of such title, kills a person while engaged in or on account of the performance of investigative, inspection, or law enforcement functions added to such section 1114 by paragraph (1) of this subsection, and who would otherwise be subject to the penalty provisions of such section 1111, shall be punished by imprisonment for any term of years or for life.

(i) Any employer who violates any of the posting requirements, as prescribed under the provisions of this Act, shall be assessed a civil penalty of up to $1,000 for each violation.

(j) The Commission shall have authority to assess all civil penalties provided in this section, giving due consideration to the appropriateness of the penalty with respect to the size of the business of the employer being charged, the gravity of the violation, the good faith of the employer, and the history of previous violations.

(k) For purposes of this section, a serious violation shall be deemed to exist in a place of employment if there is a substantial probability that death or serious physical harm could result from a condition which exists, or from one or more practices, means, methods, operations, or processes which have been adopted or are in use, in such place of employment unless the employer did not, and could not with the exercise of reasonable diligence, know of the presence of the violation.

(l) Civil penalties owed under this Act shall be paid to the Secretary for deposit into the Treasury of the United States and shall accrue to the United States and may be recovered in a civil action in the name of the United States brought in the United States district court for the district where the violation is alleged to have occurred or where the employer has its principal office.

STATE JURISDICTION AND STATE PLANS

SEC. 18. (a) Nothing in this Act shall prevent any State agency or court from asserting jurisdiction under State law over any occupational safety or health issue with respect to which no standard is in effect under section 6.

(b) Any State which, at any time, desires to assume responsibility for development and enforcement therein of occupational safety and health standards relating to any occupational safety or health issue with respect to which a Federal standard has been promulgated under section 6 shall submit a State plan for the development of such standards and their enforcement.

(c) The Secretary shall approve the plan submitted by a State under subsection (b), or any modification thereof, if such plan in his judgment—

(1) designates a State agency or agencies as the agency or agencies responsible for administering the plan throughout the State,

(2) provides for the development and enforcement of safety and health standards relating to one or more safety or health issues, which standards (and the enforcement of which standards) are or will be at least as effective in providing safe and healthful employment and places of employment as the standards promulgated under section 6 which relate to the same issues, and which standards, when applicable to products which are distributed or used in interstate commerce, are required by compelling local conditions and do not unduly burden interstate commerce,

(3) provides for a right of entry and inspection of all workplaces subject to the Act which is at least as effective as that provided in section 8, and includes a prohibition on advance notice of inspections,

(4) contains satisfactory assurances that such agency or agencies have or will have the legal authority and qualified personnel necessary for the enforcement of such standards,

(5) gives satisfactory assurances that such State will devote adequate funds to the administration and enforcement of such standards,

(6) contains satisfactory assurances that such State will, to the extent permitted by its law, establish and maintain an effective and comprehensive occupational safety and health program applicable to all employees of public agencies of the State and its political subdivisions, which program is as effective as the standards contained in an approved plan,

(7) requires employers in the State to make reports to the Secretary in the same manner and to the same extent as if the plan were not in effect, and

(8) provides that the State agency will make such reports to the Secretary in such form and containing such information, as the Secretary shall from time to time require.

(d) If the Secretary rejects a plan submitted under subsection (b), he shall afford the State submitting the plan due notice and opportunity for a hearing before so doing.

Notice of hearing.

(e) After the Secretary approves a State plan submitted under subsection (b), he may, but shall not be required to, exercise his authority under sections 8, 9, 10, 13, and 17 with respect to comparable standards promulgated under section 6, for the period specified in the next sentence. The Secretary may exercise the authority referred to above until he determines, on the basis of actual operations under the

State plan, that the criteria set forth in subsection (c) are being applied, but he shall not make such determination for at least three years after the plan's approval under subsection (c). Upon making the determination referred to in the preceding sentence, the provisions of sections 5(a)(2), 8 (except for the purpose of carrying out subsection (f) of this section), 9, 10, 13, and 17, and standards promulgated under section 6 of this Act, shall not apply with respect to any occupational safety or health issues covered under the plan, but the Secretary may retain jurisdiction under the above provisions in any proceeding commenced under section 9 or 10 before the date of determination.

Continuing evaluation.

(f) The Secretary shall, on the basis of reports submitted by the State agency and his own inspections make a continuing evaluation of the manner in which each State having a plan approved under this section is carrying out such plan. Whenever the Secretary finds, after affording due notice and opportunity for a hearing, that in the administration of the State plan there is a failure to comply substantially with any provision of the State plan (or any assurance contained therein), he shall notify the State agency of his withdrawal of approval of such plan and upon receipt of such notice such plan shall cease to be in effect, but the State may retain jurisdiction in any case commenced before the withdrawal of the plan in order to enforce standards under the plan whenever the issues involved do not relate to the reasons for the withdrawal of the plan.

Plan rejection, review.

(g) The State may obtain a review of a decision of the Secretary withdrawing approval of or rejecting its plan by the United States court of appeals for the circuit in which the State is located by filing in such court within thirty days following receipt of notice of such decision a petition to modify or set aside in whole or in part the action of the Secretary. A copy of such petition shall forthwith be served upon the Secretary, and thereupon the Secretary shall certify and file in the court the record upon which the decision complained of was issued as provided in section 2112 of title 28, United States Code. Unless the court finds that the Secretary's decision in rejecting a proposed State plan or withdrawing his approval of such a plan is not supported by substantial evidence the court shall affirm the Secretary's decision. The judgment of the court shall be subject to review by the Supreme Court of the United States upon certiorari or certification as provided in section 1254 of title 28, United States Code.

72 Stat. 941; 80 Stat. 1323.

62 Stat. 928.

(h) The Secretary may enter into an agreement with a State under which the State will be permitted to continue to enforce one or more occupational health and safety standards in effect in such State until final action is taken by the Secretary with respect to a plan submitted by a State under subsection (b) of this section, or two years from the date of enactment of this Act, whichever is earlier.

FEDERAL AGENCY SAFETY PROGRAMS AND RESPONSIBILITIES

SEC. 19. (a) It shall be the responsibility of the head of each Federal agency to establish and maintain an effective and comprehensive occupational safety and health program which is consistent with the standards promulgated under section 6. The head of each agency shall (after consultation with representatives of the employees thereof)—

(1) provide safe and healthful places and conditions of employment, consistent with the standards set under section 6;

(2) acquire, maintain, and require the use of safety equipment, personal protective equipment, and devices reasonably necessary to protect employees;

(3) keep adequate records of all occupational accidents and illnesses for proper evaluation and necessary corrective action; Recordkeeping.

(4) consult with the Secretary with regard to the adequacy as to form and content of records kept pursuant to subsection (a)(3) of this section; and

(5) make an annual report to the Secretary with respect to occupational accidents and injuries and the agency's program under this section. Such report shall include any report submitted under section 7902(e)(2) of title 5, United States Code. Annual report. 80 Stat. 530.

(b) The Secretary shall report to the President a summary or digest of reports submitted to him under subsection (a)(5) of this section, together with his evaluations of and recommendations derived from such reports. The President shall transmit annually to the Senate and the House of Representatives a report of the activities of Federal agencies under this section. Report to President. Report to Congress.

(c) Section 7902(c)(1) of title 5, United States Code, is amended by inserting after "agencies" the following: "and of labor organizations representing employees".

(d) The Secretary shall have access to records and reports kept and filed by Federal agencies pursuant to subsections (a) (3) and (5) of this section unless those records and reports are specifically required by Executive order to be kept secret in the interest of the national defense or foreign policy, in which case the Secretary shall have access to such information as will not jeopardize national defense or foreign policy. Records, etc.; availability.

RESEARCH AND RELATED ACTIVITIES

SEC. 20. (a)(1) The Secretary of Health, Education, and Welfare, after consultation with the Secretary and with other appropriate Federal departments or agencies, shall conduct (directly or by grants or contracts) research, experiments, and demonstrations relating to occupational safety and health, including studies of psychological factors involved, and relating to innovative methods, techniques, and approaches for dealing with occupational safety and health problems.

(2) The Secretary of Health, Education, and Welfare shall from time to time consult with the Secretary in order to develop specific plans for such research, demonstrations, and experiments as are necessary to produce criteria, including criteria identifying toxic substances, enabling the Secretary to meet his responsibility for the formulation of safety and health standards under this Act; and the Secretary of Health, Education, and Welfare, on the basis of such research, demonstrations, and experiments and any other information available to him, shall develop and publish at least annually such criteria as will effectuate the purposes of this Act.

(3) The Secretary of Health, Education, and Welfare, on the basis of such research, demonstrations, and experiments, and any other information available to him, shall develop criteria dealing with toxic materials and harmful physical agents and substances which will describe exposure levels that . re safe for various periods of employment, including but not limited to the exposure levels at which no employee will suffer impaired health or functional capacities or diminished life expectancy as a result of his work experience.

(4) The Secretary of Health, Education, and Welfare shall also conduct special research, experiments, and demonstrations relating to occupational safety and health as are necessary to explore new problems, including those created by new technology in occupational safety and health, which may require ameliorative action beyond that

which is otherwise provided for in the operating provisions of this Act. The Secretary of Health, Education, and Welfare shall also conduct research into the motivational and behavioral factors relating to the field of occupational safety and health.

Toxic substances, records.

(5) The Secretary of Health, Education, and Welfare, in order to comply with his responsibilities under paragraph (2), and in order to develop needed information regarding potentially toxic substances or harmful physical agents, may prescribe regulations requiring employers to measure, record, and make reports on the exposure of employees to substances or physical agents which the Secretary of Health, Education, and Welfare reasonably believes may endanger the health or safety of employees. The Secretary of Health, Education, and Welfare also is authorized to establish such programs of medical examinations and tests as may be necessary for determining the incidence of occupational illnesses and the susceptibility of employees to such illnesses. Nothing in this or any other provision of this Act shall be deemed to authorize or require medical examination, immunization, or treatment for those who object thereto on religious grounds, except where such is necessary for the protection of the health or safety of others. Upon the request of any employer who is required to measure and record exposure of employees to substances or physical agents as provided under this subsection, the Secretary of Health, Education, and Welfare shall furnish full financial or other assistance to such employer for the purpose of defraying any additional expense incurred by him in carrying out the measuring and recording as provided in this subsection.

Medical examinations.

Toxic substances, publication.

(6) The Secretary of Health, Education, and Welfare shall publish within six months of enactment of this Act and thereafter as needed but at least annually a list of all known toxic substances by generic family or other useful grouping, and the concentrations at which such toxicity is known to occur. He shall determine following a written request by any employer or authorized representative of employees, specifying with reasonable particularity the grounds on which the request is made, whether any substance normally found in the place of employment has potentially toxic effects in such concentrations as used or found; and shall submit such determination both to employers and affected employees as soon as possible. If the Secretary of Health, Education, and Welfare determines that any substance is potentially toxic at the concentrations in which it is used or found in a place of employment, and such substance is not covered by an occupational safety or health standard promulgated under section 6, the Secretary of Health, Education, and Welfare shall immediately submit such determination to the Secretary, together with all pertinent criteria.

Annual studies.

(7) Within two years of enactment of this Act, and annually thereafter the Secretary of Health, Education, and Welfare shall conduct and publish industrywide studies of the effect of chronic or low-level exposure to industrial materials, processes, and stresses on the potential for illness, disease, or loss of functional capacity in aging adults.

Inspections.

(b) The Secretary of Health, Education, and Welfare is authorized to make inspections and question employers and employees as provided in section 8 of this Act in order to carry out his functions and responsibilities under this section.

Contract authority.

(c) The Secretary is authorized to enter into contracts, agreements, or other arrangements with appropriate public agencies or private organizations for the purpose of conducting studies relating to his responsibilities under this Act. In carrying out his responsibilities under this subsection, the Secretary shall cooperate with the Secretary of Health, Education, and Welfare in order to avoid any duplication of efforts under this section.

(d) Information obtained by the Secretary and the Secretary of Health, Education, and Welfare under this section shall be disseminated by the Secretary to employers and employees and organizations thereof.

Delegation of functions.

(e) The functions of the Secretary of Health, Education, and Welfare under this Act shall, to the extent feasible, be delegated to the Director of the National Institute for Occupational Safety and Health established by section 22 of this Act.

TRAINING AND EMPLOYEE EDUCATION

SEC. 21. (a) The Secretary of Health, Education, and Welfare, after consultation with the Secretary and with other appropriate Federal departments and agencies, shall conduct, directly or by grants or contracts (1) education programs to provide an adequate supply of qualified personnel to carry out the purposes of this Act, and (2) informational programs on the importance of and proper use of adequate safety and health equipment.

(b) The Secretary is also authorized to conduct, directly or by grants or contracts, short-term training of personnel engaged in work related to his responsibilities under this Act.

(c) The Secretary, in consultation with the Secretary of Health, Education, and Welfare, shall (1) provide for the establishment and supervision of programs for the education and training of employers and employees in the recognition, avoidance, and prevention of unsafe or unhealthful working conditions in employments covered by this Act, and (2) consult with and advise employers and employees, and organizations representing employers and employees as to effective means of preventing occupational injuries and illnesses.

NATIONAL INSTITUTE FOR OCCUPATIONAL SAFETY AND HEALTH

Establishment.

SEC. 22. (a) It is the purpose of this section to establish a National Institute for Occupational Safety and Health in the Department of Health, Education, and Welfare in order to carry out the policy set forth in section 2 of this Act and to perform the functions of the Secretary of Health, Education, and Welfare under sections 20 and 21 of this Act.

Director, appointment, term.

(b) There is hereby established in the Department of Health, Education, and Welfare a National Institute for Occupational Safety and Health. The Institute shall be headed by a Director who shall be appointed by the Secretary of Health, Education, and Welfare, and who shall serve for a term of six years unless previously removed by the Secretary of Health, Education, and Welfare.

(c) The Institute is authorized to—

(1) develop and establish recommended occupational safety and health standards; and

(2) perform all functions of the Secretary of Health, Education, and Welfare under sections 20 and 21 of this Act.

(d) Upon his own initiative, or upon the request of the Secretary or the Secretary of Health, Education, and Welfare, the Director is authorized (1) to conduct such research and experimental programs as he determines are necessary for the development of criteria for new and improved occupational safety and health standards, and (2) after consideration of the results of such research and experimental programs make recommendations concerning new or improved occupational safety and health standards. Any occupational safety and health standard recommended pursuant to this section shall immediately be forwarded to the Secretary of Labor, and to the Secretary of Health, Education, and Welfare.

(e) In addition to any authority vested in the Institute by other provisions of this section, the Director, in carrying out the functions of the Institute, is authorized to—

(1) prescribe such regulations as he deems necessary governing the manner in which its functions shall be carried out;

(2) receive money and other property donated, bequeathed, or devised, without condition or restriction other than that it be used for the purposes of the Institute and to use, sell, or otherwise dispose of such property for the purpose of carrying out its functions;

(3) receive (and use, sell, or otherwise dispose of, in accordance with paragraph (2)), money and other property donated, bequeathed, or devised to the Institute with a condition or restriction, including a condition that the Institute use other funds of the Institute for the purposes of the gift;

(4) in accordance with the civil service laws, appoint and fix the compensation of such personnel as may be necessary to carry out the provisions of this section;

(5) obtain the services of experts and consultants in accordance with the provisions of section 3109 of title 5, United States Code;

80 Stat. 416.

(6) accept and utilize the services of voluntary and noncompensated personnel and reimburse them for travel expenses, including per diem, as authorized by section 5703 of title 5, United States Code;

83 Stat. 190.

(7) enter into contracts, grants or other arrangements, or modifications thereof to carry out the provisions of this section, and such contracts or modifications thereof may be entered into without performance or other bonds, and without regard to section 3709 of the Revised Statutes, as amended (41 U.S.C. 5), or any other provision of law relating to competitive bidding;

(8) make advance, progress, and other payments which the Director deems necessary under this title without regard to the provisions of section 3648 of the Revised Statutes, as amended (31 U.S.C. 529); and

(9) make other necessary expenditures.

Annual report to HEW, President, and Congress.

(f) The Director shall submit to the Secretary of Health, Education, and Welfare, to the President, and to the Congress an annual report of the operations of the Institute under this Act, which shall include a detailed statement of all private and public funds received and expended by it, and such recommendations as he deems appropriate.

GRANTS TO THE STATES

SEC. 23. (a) The Secretary is authorized, during the fiscal year ending June 30, 1971, and the two succeeding fiscal years, to make grants to the States which have designated a State agency under section 18 to assist them—

(1) in identifying their needs and responsibilities in the area of occupational safety and health,

(2) in developing State plans under section 18, or

(3) in developing plans for—

(A) establishing systems for the collection of information concerning the nature and frequency of occupational injuries and diseases;

(B) increasing the expertise and enforcement capabilities of their personnel engaged in occupational safety and health programs; or

(C) otherwise improving the administration and enforcement of State occupational safety and health laws, including standards thereunder, consistent with the objectives of this Act.

(b) The Secretary is authorized, during the fiscal year ending June 30, 1971, and the two succeeding fiscal years, to make grants to the States for experimental and demonstration projects consistent with the objectives set forth in subsection (a) of this section.

(c) The Governor of the State shall designate the appropriate State agency for receipt of any grant made by the Secretary under this section.

(d) Any State agency designated by the Governor of the State desiring a grant under this section shall submit an application therefor to the Secretary.

(e) The Secretary shall review the application, and shall, after consultation with the Secretary of Health, Education, and Welfare, approve or reject such application.

(f) The Federal share for each State grant under subsection (a) or (b) of this section may not exceed 90 per centum of the total cost of the application. In the event the Federal share for all States under either such subsection is not the same, the differences among the States shall be established on the basis of objective criteria.

(g) The Secretary is authorized to make grants to the States to assist them in administering and enforcing programs for occupational safety and health contained in State plans approved by the Secretary pursuant to section 18 of this Act. The Federal share for each State grant under this subsection may not exceed 50 per centum of the total cost to the State of such a program. The last sentence of subsection (f) shall be applicable in determining the Federal share under this subsection.

(h) Prior to June 30, 1973, the Secretary shall, after consultation with the Secretary of Health, Education, and Welfare, transmit a report to the President and to the Congress, describing the experience under the grant programs authorized by this section and making any recommendations he may deem appropriate.

Report to President and Congress.

STATISTICS

SEC. 24. (a) In order to further the purposes of this Act, the Secretary, in consultation with the Secretary of Health, Education, and Welfare, shall develop and maintain an effective program of collection, compilation, and analysis of occupational safety and health statistics. Such program may cover all employments whether or not subject to any other provisions of this Act but shall not cover employments excluded by section 4 of the Act. The Secretary shall compile accurate statistics on work injuries and illnesses which shall include all disabling, serious, or significant injuries and illnesses, whether or not involving loss of time from work, other than minor injuries requiring only first aid treatment and which do not involve medical treatment, loss of consciousness, restriction of work or motion, or transfer to another job.

(b) To carry out his duties under subsection (a) of this section, the Secretary may—

(1) promote, encourage, or directly engage in programs of studies, information and communication concerning occupational safety and health statistics;

(2) make grants to States or political subdivisions thereof in order to assist them in developing and administering programs dealing with occupational safety and health statistics; and

(3) arrange, through grants or contracts, for the conduct of such research and investigations as give promise of furthering the objectives of this section.

(c) The Federal share for each grant under subsection (b) of this section may be up to 50 per centum of the State's total cost.

(d) The Secretary may, with the consent of any State or political subdivision thereof, accept and use the services, facilities, and employees of the agencies of such State or political subdivision, with or without reimbursement, in order to assist him in carrying out his functions under this section.

Reports.

(e) On the basis of the records made and kept pursuant to section 8(c) of this Act, employers shall file such reports with the Secretary as he shall prescribe by regulation, as necessary to carry out his functions under this Act.

(f) Agreements between the Department of Labor and States pertaining to the collection of occupational safety and health statistics already in effect on the effective date of this Act shall remain in effect until superseded by grants or contracts made under this Act.

AUDITS

SEC. 25. (a) Each recipient of a grant under this Act shall keep such records as the Secretary or the Secretary of Health, Education, and Welfare shall prescribe, including records which fully disclose the amount and disposition by such recipient of the proceeds of such grant, the total cost of the project or undertaking in connection with which such grant is made or used, and the amount of that portion of the cost of the project or undertaking supplied by other sources, and such other records as will facilitate an effective audit.

(b) The Secretary or the Secretary of Health, Education, and Welfare, and the Comptroller General of the United States, or any of their duly authorized representatives, shall have access for the purpose of audit and examination to any books, documents, papers, and records of the recipients of any grant under this Act that are pertinent to any such grant.

ANNUAL REPORT

SEC. 26. Within one hundred and twenty days following the convening of each regular session of each Congress, the Secretary and the Secretary of Health, Education, and Welfare shall each prepare and submit to the President for transmittal to the Congress a report upon the subject matter of this Act, the progress toward achievement of the purpose of this Act, the needs and requirements in the field of occupational safety and health, and any other relevant information. Such reports shall include information regarding occupational safety and health standards, and criteria for such standards, developed during the preceding year; evaluation of standards and criteria previously developed under this Act, defining areas of emphasis for new criteria and standards; an evaluation of the degree of observance of applicable occupational safety and health standards, and a summary of inspection and enforcement activity undertaken; analysis and evaluation of research activities for which results have been obtained under governmental and nongovernmental sponsorship; an analysis of major occupational diseases; evaluation of available control and measurement technology for hazards for which standards or criteria have been developed during the preceding year; description of cooperative efforts undertaken between Government agencies and other interested parties in the implementation of this Act during the preceding year; a progress report on the development of an adequate supply of trained manpower in the field of occupational safety and health, including estimates of future needs and the efforts being made by Government and others to meet those needs; listing of all toxic substances in industrial usage for which labeling requirements, criteria, or standards have not yet been established; and such recommendations for additional legislation as are deemed necessary to protect the safety and health of the worker and improve the administration of this Act.

NATIONAL COMMISSION ON STATE WORKMEN'S COMPENSATION LAWS

SEC. 27. (a) (1) The Congress hereby finds and declares that—

(A) the vast majority of American workers, and their families, are dependent on workmen's compensation for their basic economic security in the event such workers suffer disabling injury or death in the course of their employment; and that the full protection of American workers from job-related injury or death requires an adequate, prompt, and equitable system of workmen's compensation as well as an effective program of occupational health and safety regulation; and

(B) in recent years serious questions have been raised concerning the fairness and adequacy of present workmen's compensation laws in the light of the growth of the economy, the changing nature of the labor force, increases in medical knowledge, changes in the hazards associated with various types of employment, new technology creating new risks to health and safety, and increases in the general level of wages and the cost of living.

(2) The purpose of this section is to authorize an effective study and objective evaluation of State workmen's compensation laws in order to determine if such laws provide an adequate, prompt, and equitable system of compensation for injury or death arising out of or in the course of employment.

Establishment.

(b) There is hereby established a National Commission on State Workmen's Compensation Laws.

Membership.

(c) (1) The Workmen's Compensation Commission shall be composed of fifteen members to be appointed by the President from among members of State workmen's compensation boards, representatives of insurance carriers, business, labor, members of the medical profession having experience in industrial medicine or in workmen's compensation cases, educators having special expertise in the field of workmen's compensation, and representatives of the general public. The Secretary, the Secretary of Commerce, and the Secretary of Health, Education, and Welfare shall be ex officio members of the Workmen's Compensation Commission:

(2) Any vacancy in the Workmen's Compensation Commission shall not affect its powers.

(3) The President shall designate one of the members to serve as Chairman and one to serve as Vice Chairman of the Workmen's Compensation Commission.

Quorum.

(4) Eight members of the Workmen's Compensation Commission shall constitute a quorum.

Study.

(d)(1) The Workmen's Compensation Commission shall undertake a comprehensive study and evaluation of State workmen's compensation laws in order to determine if such laws provide an adequate, prompt, and equitable system of compensation. Such study and evaluation shall include, without being limited to, the following subjects: (A) the amount and duration of permanent and temporary disability benefits and the criteria for determining the maximum limitations thereon, (B) the amount and duration of medical benefits and provisions insuring adequate medical care and free choice of physician, (C) the extent of coverage of workers, including exemptions based on numbers or type of employment, (D) standards for determining which injuries or diseases should be deemed compensable, (E) rehabilitation, (F) coverage under second or subsequent injury funds, (G) time limits on filing claims, (H) waiting periods, (I) compulsory or elective coverage, (J) administration, (K) legal expenses, (L) the feasibility and desirability of a uniform system of reporting information concerning job-related injuries and diseases and the operation of workmen's compensation laws, (M) the resolution of conflict of laws, extraterritoriality and similar problems arising from claims with multistate aspects, (N) the extent to which private insurance carriers are excluded from supplying workmen's compensation coverage and the desirability of such exclusionary practices, to the extent they are found to exist, (O) the relationship between workmen's compensation on the one hand, and old-age, disability, and survivors insurance and other types of insurance, public or private, on the other hand, (P) methods of implementing the recommendations of the Commission.

Report to President and Congress.

(2) The Workmen's Compensation Commission shall transmit to the President and to the Congress not later than July 31, 1972, a final report containing a detailed statement of the findings and conclusions of the Commission, together with such recommendations as it deems advisable.

Hearings.

(e)(1) The Workmen's Compensation Commission or, on the authorization of the Workmen's Compensation Commission, any subcommittee or members thereof, may, for the purpose of carrying out the provisions of this title, hold such hearings, take such testimony, and sit and act at such times and places as the Workmen's Compensation Commission deems advisable. Any member authorized by the Workmen's Compensation Commission may administer oaths or affirmations to witnesses appearing before the Workmen's Compensation Commission or any subcommittee or members thereof.

(2) Each department, agency, and instrumentality of the executive branch of the Government, including independent agencies, is authorized and directed to furnish to the Workmen's Compensation Commission, upon request made by the Chairman or Vice Chairman, such information as the Workmen's Compensation Commission deems necessary to carry out its functions under this section.

(f) Subject to such rules and regulations as may be adopted by the Workmen's Compensation Commission, the Chairman shall have the power to—

(1) appoint and fix the compensation of an executive director, and such additional staff personnel as he deems necessary, without regard to the provisions of title 5, United States Code, governing appointments in the competitive service, and without regard to the provisions of chapter 51 and subchapter III of chapter 53 of such title relating to classification and General Schedule

80 Stat. 378.
5 USC 101.

5 USC 5101, 5331.

pay rates, but at rates not in excess of the maximum rate for GS-18 of the General Schedule under section 5332 of such title, and

Ante, p. 198-1.

(2) procure temporary and intermittent services to the same extent as is authorized by section 3109 of title 5, United States Code.

80 Stat. 416.

Contract authorization.

(g) The Workmen's Compensation Commission is authorized to enter into contracts with Federal or State agencies, private firms, institutions, and individuals for the conduct of research or surveys, the preparation of reports, and other activities necessary to the discharge of its duties.

Compensation; travel expenses.

(h) Members of the Workmen's Compensation Commission shall receive compensation for each day they are engaged in the performance of their duties as members of the Workmen's Compensation Commission at the daily rate prescribed for GS-18 under section 5332 of title 5, United States Code, and shall be entitled to reimbursement for travel, subsistence, and other necessary expenses incurred by them in the performance of their duties as members of the Workmen's Compensation Commission.

Appropriation.

(i) There are hereby authorized to be appropriated such sums as may be necessary to carry out the provisions of this section.

Termination.

(j) On the ninetieth day after the date of submission of its final report to the President, the Workmen's Compensation Commission shall cease to exist.

ECONOMIC ASSISTANCE TO SMALL BUSINESSES

72 Stat. 387; 83 Stat. 802. 15 USC 636.

SEC. 28. (a) Section 7(b) of the Small Business Act, as amended, is amended—

(1) by striking out the period at the end of "paragraph (5)" and inserting in lieu thereof "; and"; and

(2) by adding after paragraph (5) a new paragraph as follows:

"(6) to make such loans (either directly or in cooperation with banks or other lending institutions through agreements to participate on an immediate or deferred basis) as the Administration may determine to be necessary or appropriate to assist any small business concern in effecting additions to or alterations in the equipment, facilities, or methods of operation of such business in order to comply with the applicable standards promulgated pursuant to section 6 of the Occupational Safety and Health Act of 1970 or standards adopted by a State pursuant to a plan approved under section 18 of the Occupational Safety and Health Act of 1970, if the Administration determines that such concern is likely to suffer substantial economic injury without assistance under this paragraph."

(b) The third sentence of section 7(b) of he Small Business Act, as amended, is amended by striking out "or (5)" after "paragraph (3)" and inserting a comma followed by "(5) or (6)".

80 Stat. 132. 15 USC 633.

(c) Section 4(c)(1) of the Small Business Act, as amended, is amended by inserting "7(b)(6)," after "7(b)(5),".

(d) Loans may also be made or guaranteed for the purposes set forth in section 7(b)(6) of the Small Business Act, as amended, pursuant to the provisions of section 202 of the Public Works and Economic Development Act of 1965, as amended.

79 Stat. 556. 42 USC 3142.

ADDITIONAL ASSISTANT SECRETARY OF LABOR

SEC. 29. (a) Section 2 of the Act of April 17, 1946 (60 Stat. 91) as amended (29 U.S.C. 553) is amended by—

75 Stat. 338.

(1) striking out "four" in the first sentence of such section and inserting in lieu thereof "five"; and

(2) adding at the end thereof the following new sentence, "One of such Assistant Secretaries shall be an Assistant Secretary of Labor for Occupational Safety and Health.".

80 Stat. 462.

(b) Paragraph (20) of section 5315 of title 5, United States Code, is amended by striking out "(4)" and inserting in lieu thereof "(5)".

ADDITIONAL POSITIONS

SEC. 30. Section 5108(c) of title 5, United States Code, is amended by—

(1) striking out the word "and" at the end of paragraph (8);

(2) striking out the period at the end of paragraph (9) and inserting in lieu thereof a semicolon and the word "and"; and

(3) by adding immediately after paragraph (9) the following new paragraph:

"(10)(A) the Secretary of Labor, subject to the standards and procedures prescribed by this chapter, may place an additional twenty-five positions in the Department of Labor in GS-16, 17, and 18 for the purposes of carrying out his responsibilities under the Occupational Safety and Health Act of 1970;

"(B) the Occupational Safety and Health Review Commission, subject to the standards and procedures prescribed by this chapter, may place ten positions in GS-16, 17, and 18 in carrying out its functions under the Occupational Safety and Health Act of 1970."

EMERGENCY LOCATOR BEACONS

72 Stat. 775.
49 USC 1421.

SEC. 31. Section 601 of the Federal Aviation Act of 1958 is amended by inserting at the end thereof a new subsection as follows:

"EMERGENCY LOCATOR BEACONS

"(d)(1) Except with respect to aircraft described in paragraph (2) of this subsection, minimum standards pursuant to this section shall include a requirement that emergency locator beacons shall be installed—

"(A) on any fixed-wing, powered aircraft for use in air commerce the manufacture of which is completed, or which is imported into the United States, after one year following the date of enactment of this subsection; and

"(B) on any fixed-wing, powered aircraft used in air commerce after three years following such date.

"(2) The provisions of this subsection shall not apply to jet-powered aircraft; aircraft used in air transportation (other than air taxis and charter aircraft); military aircraft; aircraft used solely for training purposes not involving flights more than twenty miles from its base; and aircraft used for the aerial application of chemicals."

SEPARABILITY

SEC. 32. If any provision of this Act, or the application of such provision to any person or circumstance, shall be held invalid, the remainder of this Act, or the application of such provision to persons or circumstances other than those as to which it is held invalid, shall not be affected thereby.

APPROPRIATIONS

SEC. 33. There are authorized to be appropriated to carry out this Act for each fiscal year such sums as the Congress shall deem necessary.

EFFECTIVE DATE

SEC. 34. This Act shall take effect one hundred and twenty days after the date of its enactment.

Approved December 29, 1970.

LEGISLATIVE HISTORY:

HOUSE REPORTS: No. 91-1291 accompanying H.R. 16785 (Comm. on Education and Labor) and No. 91-1765 (Comm. of Conference).
SENATE REPORT No. 91-1282 (Comm. on Labor and Public Welfare).
CONGRESSIONAL RECORD, Vol. 116 (1970):
Oct. 13, Nov. 16, 17, considered and passed Senate.
Nov. 23, 24, considered and passed House, amended, in lieu of H.R. 16785.
Dec. 16, Senate agreed to conference report.
Dec. 17, House agreed to conference report.

☆ U. S. Government Printing Office: 1987 - 181-519 (64291)

Index

About the Author

Thomas A. Hunter is a registered engineer and president of Forensic Engineering Consultants, Inc., in Westport, Connecticut. He has 20 years of experience in engineering management, including assignments as technical director, chief engineer, director of research, and director of product safety.

Dr. Hunter is also the author of *Engineering Mechanics—Statics*, as well as many reports and technical papers. He received his Ph.D. in engineering mechanics from the University of Michigan, and has been a member of the American Society of Mechanical Engineers, American Society of Agricultural Engineers, American Society for Engineering Education, and American Society for Testing and Materials.